T0332063

2018 INTERNATIONAL EXISTING BUILDING CODE® HANDBOOK

Chris Kimball, SE, MCP, CBO

New York Chicago San Francisco Athens London
Madrid Mexico City Milan New Delhi
Singapore Sydney Toronto

Library of Congress Control Number: 2019944240

2018 International Existing Building Code® Handbook

1 2 3 4 5 6 QVS 23 22 21 20 19

ISBN 978-1-260-13478-0
MHID 1-260-13478-4

This book is printed on acid-free paper.

Sponsoring Editor
Lauren Poplawski

Editorial Supervisor
Stephen M. Smith

Production Supervisor
Lynn M. Messina

Acquisitions Coordinator
Elizabeth M. Houde

Project Manager
Tania Andrabi, Cenveo® Publisher Services

Copy Editor
Tulsi Rawat, Cenveo Publisher Services

Proofreader
Lisa McCoy

Art Director, Cover
Jeff Weeks

Composition
Cenveo Publisher Services

About the Author

Chris Kimball, SE, MCP, CBO, is Vice President of West Coast Code Consultants, Inc. (WC3), which provides third-party plan review services to jurisdictions throughout the United States. He is a licensed Structural Engineer in addition to being an ICC-certified Master Code Professional and Certified Building Official, and a Certified Fire Code Official. He obtained a bachelor's degree in civil engineering and a master's degree in structural engineering from Utah State University. Mr. Kimball has served as president of the Structural Engineers Association of Utah (SEAU) as well as president of the Beehive Chapter of the International Code Council. He has provided training classes to building officials, design professionals, and contractors alike to assist them in understanding the requirements of the adopted building codes.

About the International Code Council

The International Code Council® (ICC®) is a member-focused association dedicated to helping the building safety community and construction industry provide safe, sustainable, and affordable construction through the development of codes and standards used in the design, build, and compliance process. Most U.S. communities and many global markets choose the International Codes®. ICC Evaluation Service (ICC-ES), a subsidiary of the International Code Council, has been the industry leader in performing technical evaluations for code compliance fostering safe and sustainable design and construction.

Governmental Affairs Office: 500 New Jersey Avenue NW, 6th Floor, Washington, DC 20001-2070

Regional Offices: Birmingham, AL; Chicago, IL; Los Angeles, CA

1-888-422-7233; www.iccsafe.org

Contents

Chapter 9

Chapter 10

Preface

Communities in every state are faced with decaying, blighted, and vacant existing buildings. In the past, jurisdictions throughout the United States have utilized either the *International Building* Code® (IBC®) or *International Residential Code*® (IRC®) for evaluating existing buildings when they undergo alterations, additions, or changes in use. Both the IBC and IRC really are meant for new construction and, in many cases, are not appropriate for existing buildings.

The National Institute of Building Sciences recently issued a white paper in which the organization highlighted the need for an existing building code and other regulatory tools that will encourage revitalization. Such a tool exists in the *International Existing Building Code*® (IEBC®). The purpose of this book is to help owners, builders, contractors, design professionals, and, most important, code officials understand the requirements of the IEBC.

The IEBC creates a win-win scenario for both the building owner and the local jurisdiction. It allows building owners to perform improvements on their facility without requiring the entire building to be upgraded to the requirements of the IBC or IRC, while it also provides a means for code officials to ensure that existing buildings are safe.

Chris Kimball, SE, MCP, CBO

CHAPTER 1

INTRODUCTION

1.1 *Building Code History*

Building codes are used throughout the world to oversee the design and construction of buildings. Most of us know what a building is, but what is a "code"? Dictionary.com defines the word "code" as "any set of standards set forth and enforced by a local government agency for the protection of the public safety, health, etc." The primary purpose for a federal, state, or local jurisdiction to adopt a building code is to ensure the life safety of the public. The building code presents "minimum" standards for the design and construction of buildings. "These minimum requirements... attempt to represent society's compromise between optimum safety and economic feasibility" (Cote and Grant, 2008, 1–53). Building codes are also used for research, education, and knowledge assessment of building industry professionals such as architects, engineers, and building safety professionals. The following outlines several significant events that have led to the development of the building codes that are commonly used today.

1.1.1 King Hammurabi (1758 BC). The first written building codes were enacted by King Hammurabi almost 4000 years ago. Hammurabi was the sixth king of Babylon and ruled from 1792 to 1750 BC. The code of Hammurabi (Fig. 1.1) was written on 12 stone tablets and contains 282 laws. Numbers 228 through 233 provided the laws in relation to construction and are listed below (Heady, n.d.).

> *228:* If a builder has built a house for a man, and finished it, he shall pay him a fee of two shekels of silver, for each SAR built on.
>
> *229:* If a builder has built a house for a man, and has not made his work sound, and the house he built has fallen, and caused the death of its owner, that builder shall be put to death.

Figure 1.1 Code of Hammurabi. (*www.thegolfclub.info, n.d.*)

230: If it is the owner's son that is killed, the builder's son shall be put to death.

231: If it is the slave of the owner that is killed, the builder shall give slave for slave to the owner of the house.

232: If he has caused the loss of goods, he shall render back whatever he has destroyed. Moreover, because he did not make sound the house he built, and it fell, at his own cost he shall rebuild the house that fell.

233: If a builder has built a house for a man, and has not keyed his work, and the wall has fallen, that builder shall make that wall firm at his own expense.

Many complain about the requirements we have in today's modern building codes, but after reviewing the code at the time of King Hammurabi, it is apparent that we have come a long way.

1.1.2 Great Fire of London (1666). On September 2, 1666, a small fire in a bakery on London's Pudding Lane quickly spread and swept through London, destroying more than 13,000 houses, St. Paul's Cathedral, and hundreds of other commercial buildings and left over 200,000 people without residences or businesses (Fig. 1.2). The catastrophe was due to a combination of strong winds, closely built properties, and a warm summer that had dried out the wood and thatch used in the construction of most buildings.

A royal proclamation halted rebuilding until new regulations were put into place. The Rebuilding Act of 1667 was soon implemented and included the following building regulations (British History Online, n.d.):

- All buildings within the city were to be constructed of brick or stone, except for doors and window frames.

Figure 1.2 The Great Fire of London in 1666. (*By Lieve Verschuier.*)

- Four types of buildings were allowed:
  - Two-story buildings with a cellar (i.e., basement)
  - Three-story buildings with a cellar
  - Four-story buildings with a cellar
  - Mansions, not more than four stories (for citizens of "extraordinary quality")
- The thickness of all walls was precisely specified.
- Party walls (i.e., common or shared walls) were to be properly toothed for better joining.
- Balconies, or projected floors, are to use iron, lead slate, tile, or plaster finish materials and no wood was to be exposed.
- New fire prevention regulations included easy access to water and the beginnings of a fire hydrant system for the city.

Within 10 years the areas destroyed by the fire had been rebuilt. The street layout remained mostly the same but with bigger buildings mostly built with stone.

1.1.3 Chicago Fire of 1871 and the "Great Rebuilding." On October 8, 1871, the Great Chicago Fire began in the barn of the Patrick and Catherine O'Leary farm and quickly spread to the city's center due to strong southwestern winds (Fig. 1.3). The fire destroyed more than 4 square miles of the city and over 18,000 buildings, killed over 300 people, and left one-third of

Figure 1.3 Chicago in Flames by Currier and Ives. (*Wikimedia, 2016.*)

the city's population homeless. A summer drought and the wooden construction of the densely built buildings helped the fire to grow out of control (Schons, 2011).

The Great Rebuilding occurred soon after the fire. Since most railroad tracks had remained undamaged, aid reached the city from various parts of the United States. The fire initiated many changes: strengthening of safety rules and the city's fire department and laws that required new buildings to be made of fireproof materials. Such materials included brick, stone, marble, terracotta, and limestone. Just 20 years after the fire, the city's population had grown from 300,000 to 1 million people (Chicago Architecture Center, n.d.).

1.1.4 1905 *National Building Code*. Prior to the Great Chicago Fire of 1871, Lloyd's of London had stopped writing insurance policies in Chicago due to the careless nature of building construction. Insurance rates increased significantly throughout the United States for this same reason; however, even with the increased rates, insurance companies still suffered great losses due to uncontrolled fires and poor regulations (Cote and Grant 2008, 1–52).

The National Board of Fire Underwriters (NBFU), organized in 1866, realized that the adjustment and standardization of rates were merely a temporary solution to a serious technical problem. This group began to emphasize safe building construction, control of fire hazards, and improvements in both water supplies and fire departments. They developed guidelines for what was termed a "Class A" building. These were tall buildings constructed of concrete and steel and included additional measures to limit the risk of fire (Cote and Grant, 2008, 1–52).

In 1905, NBFU published the first edition of the *National Building Code* (NBC). The main purpose of the NBC was to minimize risks to property and building occupants. This original publication had a total of 263 pages, including the table of contents and index.

1.1.5 UBC, SBCC, and BOCA. The existence of the NBC led to the formation of several building official organizations.

By 1950, the United States had three regional code organizations, each with its own code. The Building Officials and Code Administrators International (BOCA) was traditionally used on the Northeast and Great Lakes portions of the United States. The codes from the Southern Building Code Congress International (SBCCI) were conventionally used throughout the Southern United States. The codes published by the International Conference of Building Officials (ICBO) were traditionally used in the Western United States but were enforced by municipalities as far east as Indiana (Cote and Grant, 2008, 1–63). The map in Fig. 1.4 roughly shows the areas in which each of these codes were regionally adopted.

1.1.6 The International Codes. The International Code Council® (ICC®) was created in 1994 with the purpose of developing a single set of comprehensive and coordinated building codes that would be used throughout the United States as well as internationally (Fig. 1.5). BOCA, SBCCI, and ICBO were all founding member organizations of ICC. The original family of ICC International Codes® (I-Codes®) was published in 2000 (Wikipedia, n.d.). The I-Codes are considered "model" codes and have no legal ramifications unless they have been formally adopted by a federal, state, or local authority.

Currently, ICC's complete family of codes includes the following:

- *2018 International Building Code®* (IBC®)
- *2018 International Residential Code®* (IRC®)
- *2018 International Fire Code®* (IFC®)
- *2018 International Existing Building Code®* (IEBC®)

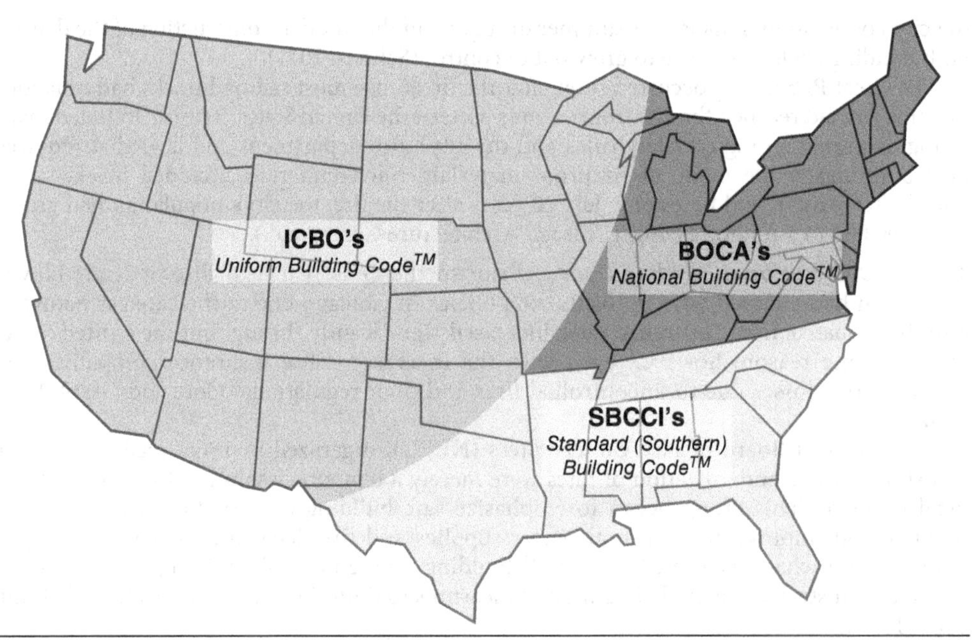

Figure 1.4 Map of pre-ICC code adoptions.

Figure 1.5 2018 ICC model codes. (*Copyright 2017. Washington, D.C.: International Code Council.*)

- *2018 International Mechanical Code*® (IMC®)
- *2018 International Plumbing Code*® (IPC®)
- *2018 International Fuel Gas Code*® (IFGC®)
- *2018 International Energy Conservation Code*® (IECC®)
- *2018 International Wildland-Urban Interface Code*® (IWUIC®)
- *2018 International Performance Code for Buildings and Facilities*® (ICCPC®)
- *2018 International Property Maintenance Code*® (IPMC®)
- *2018 International Zoning Code*® (IZC®)
- *2018 International Private Sewage Disposal Code*® (IPSDC®)
- *2018 International Swimming Pool and Spa Code*® (ISPSC®)
- *2018 International Green Construction Code*® (IgCC®)

As of May 2018, at least a portion of all U.S. states and territories, except for American Samoa, had adopted the IBC and other members of the I-Code family (International Code Council, n.d.-b). While the I-Codes are currently updated every 3 years, many states and local jurisdictions do not always adopt the current version.

1.1.7 History of the IEBC. The IEBC was first introduced in 2003 after an exhaustive effort that began in 2000. The original intent was to create a comprehensive set of regulations for existing buildings based on the requirements previously included in the codes developed by BOCA, SBCCI, and ICBO. The purpose of the IEBC is to encourage the use and reuse of existing buildings while also maintaining minimum life safety requirements. It has technical requirements that, in most cases, are more stringent than the model code under which the existing buildings were constructed, but not as stringent as the codes currently adopted for new construction. In a white paper developed by the National Institute of Building Sciences (NIBS), it was suggested that "The adoption of the IEBC… has broader community benefits by becoming a key element and driver of economic development; promoting affordable housing; allowing the reuse of existing building stock; and fostering the revitalization of older, often blighted and vacant neighborhoods" (Rodgers et al., 2017, 5).

While the IEBC was first introduced in 2003, it has taken quite some time for it to be accepted by the design and building safety communities and has been slow to be adopted for use throughout the United States. Previous to the 2015 IBC, Chapter 34 of the IBC included provisions for existing buildings. With some existing building provisions included in the IBC, and since several associations were not happy with portions of the initial IEBC requirements, the IEBC previously took a back seat to the IBC.

In the 2009 IBC, Chapter 34 specifically listed the IEBC as an alternative compliance option for existing buildings. While this was a good step in the right direction, a huge change was made with the adoption of the 2015 IBC. In the 2015 version Chapter 34 was removed entirely and Chapter 1 of the IBC now specifically lists the IEBC as a referenced code, similar to the mechanical, plumbing, fire, and energy codes. As a referenced code, and with IBC Chapter 34 removed, the only option for designers now is to cause existing buildings to comply with all of the provisions of the IBC for new structures, or to comply with the provisions of the IEBC.

With the removal of IBC Chapter 34, more states and local jurisdictions are now turning to the IEBC for the enforcement of existing buildings. As illustrated in Fig. 1.6, ICC's website currently shows that only nine U.S. states or territories have not adopted the IEBC at either the state or local level (International Code Council, n.d.-b).

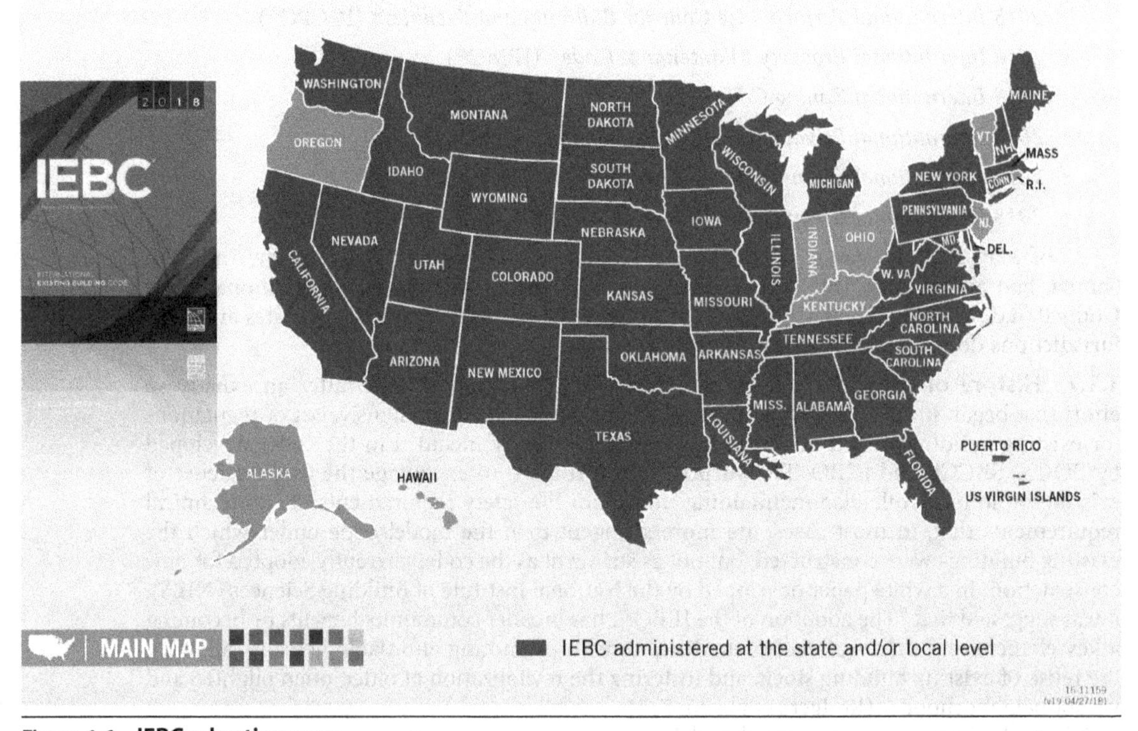

Figure 1.6 **IEBC adoption map.**

1.2 *Why the IEBC?*

The initial thought might be, "Just what we need, another code!" While it may require code officials, design professionals, and contractors to learn something new, it also provides much more flexibility for the reuse of existing buildings. The purpose of the IEBC is to preserve the existing building stock by providing a "...*reasonable* set of technical requirements and associated construction costs" (Rodgers et al., 2017, 3). Yes, it is another book, but it is only 83 pages long, not including appendices.

The IEBC provides several paths or options to the building owner and design professional, as well as to the code official. The level of compliance is dependent upon the amount of work that is being performed. As an example, repairs often require much less than what would be triggered by performing a significant remodel or making an addition to an existing building. The IEBC allows more opportunities for the design professional and code official to assist in helping more existing buildings to be safe.

In an article written in *Structure Magazine* in February 2017, Zeno Martin highlighted the need for the IEBC by describing that building codes are constantly evolving. Every 3 years a new code cycle begins, and changes are made to most design disciplines. It is not practical to require existing buildings to constantly be updated to meet the current code provisions. The IEBC therefore provides both the owner and code official a means to treat existing buildings differently (Martin et al., 2017).

1.3 *Scope*

Section 101.2 of the IEBC states that this code shall apply to the repair, alteration, change of occupancy, addition, and relocation of existing buildings. It also includes an exception allowing for certain residential occupancies to alternately comply with the provisions of the IRC.

The IRC is often referred to as a stand-alone code, but in relation to existing buildings that may not be the case. Section R102.7.1 of the IRC states that "Additions, alterations, repairs and relocations shall not cause an existing structure to become unsafe or adversely affect the performance of the building." This statement leaves it up to the code official to determine whether the addition or alteration being proposed would adversely affect the existing structure. By referring to the IEBC, the code official can reference specific provisions that are outlined therein but are not addressed in the IRC.

1.4 *Key Terms*

As with the entire I-Code family, Chapter 2 of the IEBC provides definitions for key terms that are used throughout the code. The following is a list of key words and associated definitions that will be considered throughout this book. A more detailed definition for most of these terms can be found in Section 202 of the IEBC.

- *Addition:* An extension or increase in floor area, number of stories, or height.
- *Alteration:* Any construction activity involving an existing structure other than a repair or addition.
- *Approved:* An item that is found to be acceptable to the code official.
- *Change of Occupancy:* A change in the use of the building or portion of the building.
- *Code Official:* The designated authority charged with the administration and enforcement of the code.
- *Dangerous:* Any structure, or portion thereof, that has collapsed, has partially collapsed, has moved off its foundation, lacks support, or where a significant risk of collapse exists.
- *Repair:* The reconstruction, replacement, or renewal of any part of a structure for maintenance purposes or to correct damage.
- *Structural Engineer of Record (SER):* The licensed engineer who is responsible for the structural design of the building or structure.
- *Substantial Damage:* Structures that have sustained damage for which the cost of restoration would equal or exceed 50 percent of cost of the undamaged structure.

- *Substantial Improvement:* Any repair, alteration, addition, or improvement of a structure that equals or exceeds 50 percent of the structure's cost prior to said improvement.
- *Substantial Structural Alteration:* Alterations to existing structural elements that combine to support more than 30 percent of the floor or roof area.
- *Substantial Structural Damage:* This occurs when (1) the capacity of wind and seismic resisting members are reduced by 33 percent or more, (2) structural elements that combine to support more than 30 percent of the floor or roof area are reduced by 20 percent or more, and (3) the capacity of structural elements supporting snow loads are reduced by 20 percent or more.
- *Technically Infeasible:* An alteration that is not likely to occur due to the locations of structural elements or due to site constraints, which would limit full code compliance.
- *Unsafe:* Structures that have insufficient means of egress, inadequate light and ventilation, constitute a fire hazard, include structural elements deemed "dangerous," or include other aspects that would be unsanitary or dangerous for human occupancy.
- *Work Area:* Portions of a building that are being altered or reconfigured.

1.5 *Layout*

When using the IEBC, it is very important to understand how it is laid out. Because it has multiple paths of compliance, as well as several forks in the road within each path, understanding how to navigate through the code will allow the user to find the most beneficial option. Figure 1.7 shows how the IEBC is broken down by chapter, while the following discussion provides some additional information that the user should be aware of.

Chapters	Subjects
1–2	Administrative Requirements and Definitions
3	Provisions for all Compliance Methods
4	Repairs
5	Prescriptive Compliance Method for Existing Buildings
6–12	Work Area Compliance Method for Existing Buildings
13	Performance Compliance Method for Existing Buildings
14	Relocated Buildings
15	Construction Safeguards

Figure 1.7 IEBC table "Arrangement and Format of the 2018 IEBC." (*Excerpted from the 2018 International Existing Building Code; copyright 2017. Washington, D.C.: International Code Council.*)

1.5.1 Compliance methods. The IEBC provides three main options to the designer when dealing with alterations, additions, or the change of use in relation to existing buildings. Chapter 5 outlines the "Prescriptive Compliance Method." This method is almost identical to

Chapter 34 from the 2012 IBC. Chapters 6 to 12 outline the "Work Area Compliance Method" option. Most of the benefits for using the IEBC are frequently found when using this method. The third option is covered in Chapter 13 and addresses the "Performance Compliance Method." This method is the least used and, most likely, the least understood.

1.5.2 Repairs. Chapter 6 of the IEBC governs the repair of existing buildings. This chapter is independent of the compliance methods described above that are for the alteration, addition, or change of use of existing buildings.

1.5.3 Moved buildings. Chapter 14 of the IEBC outlines the code requirements for moved buildings. Similar to Chapter 6, Repairs, this chapter is independent of the three compliance methods described above.

1.5.4 Referenced standards. Chapter 16 provides a list of the standards that are referenced throughout the IEBC. These standards are considered part of the IEBC to the extent of the reference within the code. The user should be aware of the requirements within the standard and should also ensure that the appropriate edition is considered as listed in Chapter 16.

1.5.5 Appendices. The appendices provided in the International Codes are provided to offer optional or supplemental criteria to the main code provisions, and those provided in the IEBC are no different. It is important to realize the appendices typically need to be specifically adopted by the state or local jurisdiction to be applicable, so the designer should ensure that such is the case prior to referencing them in their design.

The appendices in the IEBC provide specific guidelines for performing seismic retrofits of specific types of buildings, wind strengthening provisions for wood-framed buildings, and include supplementary accessibility guidelines for existing buildings. In addition to these appendices, the IEBC includes a resource section that provides fire-resistance rating information for older building materials and assemblies that are not outlined in Chapter 7 of the IBC.

1.6 *2018 IEBC Updates*

As with all of the International Codes, the user is able to identify changes that have been made within the code from the previous version. If an arrow (→) is shown in the margin it means that something was removed from that location of the code since the previous version. If the margin includes a single asterisk (*) it indicates that something that was previously in that location of the code has been moved to somewhere else within the same code. If a double asterisk (**) is shown it designates a location where the language has been moved to. Figure 1.8 displays items that have been relocated from the 2015 version of the IEBC to the 2018 version.

2018 LOCATION	2015 LOCATION
302.2	401.3
305	410
904.1.4	804.2.4
1201.5	1202.2
1206.1	1202.3

Figure 1.8 IEBC table "Marginal Markings." ***(Excerpted from the 2018 International Existing Building Code; copyright 2017. Washington, D.C.: International Code Council.)***

Chapter, 2018	Chapter, 2015	Title
4	6	Repairs
5	4	Prescriptive Method
6	5	Classification of Work
13	14	Performance Method
14	13	Relocated or Moved Buildings

Figure 1.9 IEBC table "Chapter Reorganization." ***(Excerpted from the 2018 International Existing Building Code; copyright 2017. Washington, D.C.: International Code Council.)***

The 2018 version of the IEBC includes several significant updates that users should be aware of. Figure 1.9 shows how the chapter numbering has been reorganized from the previous version. The following list highlights some of the significant changes that have been made:

- The accessibility provisions have been removed from the individual compliance methods and are now located in Chapter 3, under Section 305.
- The live load requirements included in the prescriptive and work area methods have been removed and are now provided in Chapter 3 to apply to all compliance methods.
- The IEBC now requires structural components that have been damaged by snow events to be repaired assuming the snow loads for new buildings per the IBC.
- The prescriptive compliance method now requires the addition of wall anchors at the roof line for reinforced concrete and masonry walls located in high-seismic regions.
- The prescriptive compliance method now requires that live loads, snow loads, wind loads, and seismic loads be checked for a change of occupancy.
- Additions to educational occupancies that would require storm shelters under the IBC for new construction now require a storm shelter to be constructed if the occupant load is increased by 50 or more.
- Carbon monoxide requirements are now included under both the prescriptive and work area methods during alterations and additions.
- Existing emergency escape and rescue openings must now comply with operational criteria that are listed in the IBC under both the prescriptive and work area methods.
- Chapter 6, addressing repairs, has been renumbered as Chapter 4. In addition, the repair provisions included in the prescriptive compliance method have been removed, making Chapter 4 the reference for all repairs.
- Chapter 13, addressing moved buildings, has been renumbered as Chapter 14. In addition, the moved structure requirements that were included in the prescriptive compliance method have been removed, making Chapter 14 the reference in all cases when a building is to be relocated.

CHAPTER 2

ADMINISTRATION

2.1 *Duties of the Building Department*

The building department is the entity charged with enforcing the requirements of the adopted codes, their referenced standards, and any applicable local ordinances. Throughout this book the term "code official" will be used. This term could apply to almost anyone working within the building department and not just the designated building official. As noted in Section 103.3 of the *International Existing Building Code*® (IEBC®), other members of the department such as permit technicians, plans examiners, and building inspectors are deemed deputies to the code official. It also encompasses officials in various disciplines of building safety such as plumbing, mechanical, energy, fire safety, accessibility, structural, and others.

The following is a listing of many of the code official duties outlined in Chapter 1 of the IEBC:

- Enforce the code
- Render interpretations of the code
- Adopt policies and procedures for code enforcement
- Receive applications for permits
- Review construction documents
- Issue permits
- Inspect the premises
- Determine if substantially improved
- Determine if substantially damaged
- Attend preliminary application meetings
- Issue notices and orders
- Perform required code inspections
- Retain official records

After reviewing these requirements, it is easy to see why deputies are required to perform the duties of the code official. Just complying with the first bullet item, enforcing the code, is a daunting task. The code encompasses so many different disciplines of construction, including architectural, structural, civil, mechanical, plumbing, electrical, energy, etc. One individual can never know it all. The code official must understand the code well enough to render interpretations and must be careful not to waive specific code requirements.

Sections 104.10 and 104.11 of the IEBC allow for the code official to accept modifications or alternate materials, designs, and methods of construction that do not meet the letter of the code. In no case can the code official allow a modification or alternate that does not meet the intent of the code. The decision made by the code official must be documented, and if desired, research reports or tests can be required to support the decision.

The code official, and any designated deputies, must carry proper identification when performing inspections. If the code official has reasonable cause to believe that a code violation exists within an existing structure, Section 104.6 of the IEBC allows for the official to enter the premises at reasonable times. If the structure is occupied, the code official must be able to present their identification. If unoccupied, reasonable accommodations should be made to locate the owner and request entry. If entry is refused in either an occupied or unoccupied structure, the code official may need to seek the assistance of law enforcement.

As long as the code official and any of their deputies are acting in good faith and without malice, they are protected from personal liability when enforcing the provisions of the code. Section 104.8.1 of the IEBC further states that if there is a criminal complaint, the jurisdiction is required to provide legal representation at no cost to the employee.

2.2 *Work Requiring a Permit*

Section 105 of the IEBC outlines when a building permit must be obtained. In general, all work involving the repair, addition, alteration, relocation, demolition, or change of occupancy to an existing building or its mechanical, electrical, and plumbing components requires a permit unless one of the following items applies:

Building Items:

- Concrete flatwork (i.e., driveways or sidewalks)
- Finish work (i.e., painting, papering, tiling, carpeting, cabinets, countertops, etc.)
- Temporary stage sets or scenery
- Shade cloth structures used for agricultural purposes
- Window awnings attached to a single-family residence
- Movable cases, counters, and partitions not over 69 in. in height

Electrical Items:

- Minor repairs or maintenance work
- Radio or television transmitting stations
- Temporary testing or servicing equipment

Gas Items:

- Portable heating appliances
- Replacement of minor part

Mechanical Items:

- Portable heating appliance
- Portable ventilation equipment
- Portable cooling unit
- Water piping within heating or cooling equipment
- Replacement of minor part
- Portable evaporative cooler
- Refrigeration system containing less than 10 lb of refrigerant

Plumbing Items:

- The stopping of leaks
- The clearing of stoppages

As noted above, almost all work within an existing building will require an application for a building permit, unless said work is of a minor nature. In addition to the exceptions noted above, Section 105.2.2 of the IEBC states that repairs are exempt from requiring a permit if they do not include any of the following items.

Repairs requiring a permit:

- The cutting of any wall, partition, or portion thereof
- The removal or cutting of any structural beam or load-bearing support
- The removal or change to a required means of egress component
- Any addition to, alteration of, replacement, or relocation of any standpipe, water supply, sewer, drainage, drain leader, gas, soil, waste, vent, or similar piping, or electric wiring
- Mechanical or other work affecting public health or general safety

Section 105.2.1 of the IEBC allows for emergency repairs, which could include some of the items noted above, to be performed if a permit application is submitted to the code official by the next business day. A good example of when an emergency repair may be needed is shown in Fig. 2.1 (Laslo, 2011). It is quite surprising how often cars, or other vehicles, crash into buildings. A quick Google search would lead to thousands of results. If the damage occurs to a commercial building, such as the Kentucky Fried Chicken restaurant shown in Fig. 2.1, it is imperative that the repairs are made immediately to minimize the danger to its workers and to the public, as well as the loss of revenue that will occur while it is closed to the public.

It is also important to note that building permits are not required for the installation, alteration, or repair of generation, transmission, distribution of utilities, or metering equipment that is under the control of public service agencies. While a building permit issued by a local jurisdiction is not necessarily required for these types of projects, there are other regulatory authorities that oversee public utilities.

Figure 2.1 **Bus crash into KFC.**

2.3 *Permit Submittals*

Almost all types of work requiring a permit also require that certain items be submitted to the local jurisdiction. The purpose of these submittals is to provide sufficient clarity for how the work is proposed to be constructed and to show that the minimum requirements of the code will be met. Many municipalities do not require submittals for work such as the installation of a new water heater. For these types of projects, the code official will ensure that they have been installed correctly during inspection.

Every jurisdiction is a bit different, and as such the permit applicant should review the local jurisdiction's website to see what submittals are required. If this information is not provided on their website, a call should be made, as an incomplete submittal will slow down the review process and delay the issuance of a building permit. At the same time, municipalities should do all they can to make this information readily available to permit applicants on the municipality's website or make clear forms or handouts available. Figure 2.2 provides an example of a submittal checklist for commercial projects within the City of Mercer Island, Washington (City of Mercer Island, 2013).

As outlined in this checklist, there could be many items that are required as part of the permit application submittals. The following sections present a brief description for several of the items that could be required as part of a permit application for construction activity affecting an existing building.

2.3.1 Construction documents (IEBC § 106.2). Section 202 of the IBC defines the construction documents as "Written, graphic, and pictorial documents prepared or assembled for describing the design, location, and physical characteristics of the elements of the project necessary for obtaining a building permit." The construction documents are those submittals that clearly show all of the work that is to occur on the project. In most cases the construction documents are simply considered to be the construction drawings, or plans; however, if project specifications exist, they should also be considered part of the construction documents.

Section 106.2 of the IEBC states that the construction documents must be "...of sufficient clarity to indicate the location, nature, and extent of the work." It also requires the documents to clearly show conformance with the provisions laid out in the code, as well as any local laws and ordinances. Section 106.2 also contains several subsections noting that the construction documents must clearly list the requirements for required fire protection systems; all required means of egress elements, clarifying the exterior wall envelope requirements; and the construction of exterior balconies or elevated walking surfaces.

Section 106.2.6 of the IEBC requires that the construction documents also include a site plan. The site plan must be drawn to scale and shall show the size and location of all existing and new structures on the site. The distances between these structures and the lot lines must be clearly shown in addition to the proposed finished grades.

In addition to what is required by the IEBC, Section 1603.1 of the IBC requires that certain structural information be shown on the construction drawings. This includes "...the size, section, and relative locations of structural members with floor levels, column centers, and offsets dimensioned." The design loads, snow data, wind data, earthquake data, and flood data considered in the design must also be noted.

2.3.2 Geotechnical investigations. While the IEBC does not specifically require a geotechnical report, some alterations, additions, or moved buildings will require such a report. Remember that the IEBC is actually a referenced code in the IBC per Section 101.4.7. As such, the requirements within the IBC apply unless there is a specific provision within the IEBC that would not require it for existing buildings.

City of Mercer Island
9611 SE 36th Street • Mercer Island, WA 98040-3732
PHONE (206) 275-7605 • FAX (206) 275-7726
www.mercergov.org

Submittal Checklist for Commercial Projects

The following documents are required to be submitted at the Pre-Application Meeting. Unless noted otherwise, **2 copies** of plans and documents are required. A packet with all of these required documents can be picked up at Mercer Island City Hall. Most of these documents can also be downloaded from the city website: **www.mercergov.org/**

		Submittal	N/A	Staff Use
1	**Building Permit Application** complete with site address and parcel number; owner's name, address and phone number; contractor's name, address, phone number, state contractor's license number, and Mercer Island business license number. **(1 copy)**			
2	**Geotechnical Report** completed by a Geotechnical Engineer. **(2 copies)**			
3	**Energy & Lighting Calculations or Worksheet. (1 copy)**			
4	**Structural Calculations. (1 copy)**			
5	**Storm Drainage Report/ Hydraulic Calculations**			
6	**Water Meter Sizing Worksheet (1 copy)**			
7	**Fire Area Square Footage Calculations Worksheet (1 copy)**			
8	**Topographic Survey** stamped, signed and dated by the surveyor, if required. **(1 copy)**			
9	**2 Small Site Plans.** On 8-1/2" X 11" paper. May be a reduction of the full size site plans.			
10	**2 Complete Sets of Plans** drawn at a minimum scale of ¼" = 1' showing conformance to applicable building codes and including structural notes and material specifications. Include items a. through k. below plus any other sheets as necessary.			
a.	**Site Plans based on a Topographic Survey** (min. scale 1"=20') including the following information:			
	• legal description of property, unless a separate topographic survey is required for this project;			
	• north arrow and scale;			
	• all property lines and easements with dimensions;			
	• accurate existing and proposed topography at 2 foot maximum contour intervals;			
	• all setbacks from property lines, shorelines, and watercourses with dimensions;			
	• all critical areas and buffers (critical slopes, shorelines, streams, wetlands, geological hazard areas);			
	• centerline of adjacent streets/alleys, street names and whether street is public or private;			
	• location and dimensions of all *existing* buildings/structures including retaining walls and rockeries clearly marked whether they will remain or be demolished;			
	• location and dimensions of all *proposed* buildings/structures including retaining walls and rockeries;			
	• on-site parking and driveways; and			
	• improvements in the city right-of-way, including driveways, utilities and landscaping.			
b.	**Foundation Plans.** Show dimensions, anchor bolts, holdowns, vent size and location, size and location of crawl space access, and connection details (especially when connecting new foundation to existing).			
c.	**Floor Plans.** Show all dimensions, room names, and window sizes (with egress windows and safety glass clearly labeled). Include north arrow on all plan drawings.			
d.	**Structural Framing Plans.** Show all structural details for roof systems, floor systems, and deck framing.			
e.	**Cross Sections.** Show at least one full cross section taken at a location which describes the building best at a min. ¼"=1' or larger; at least one dimensioned section of each different foundation condition if not shown elsewhere; and at least one typical wall section fully detailed to show basic construction materials to be used at ½"= 1' scale min.			
e.	**Cross Sections.** Show at least one full cross section taken at a location which describes the building best at a min. ½" =1' or ¾"=1' scale; at least one dimensioned section of each different foundation condition if not shown elsewhere; and at least one typical wall section fully detailed to show basic construction materials to be used at ¼" = 1' scale min.			
f.	**Elevations.** Show one elevation view for each side of new construction, plus any needed to fully describe additions. Include the location of existing grade, average building elevation, and maximum building height allowed.			
g.	**Erosion Control Plans.** May be incorporated on the site plan or in the Stormwater/Utility Plan. Show location of all temporary erosion and sediment control measures.			
h.	**Site Restoration Plans.** *(If authorized to work in a critical area).* If critical areas (steep slopes, streams, shoreline, wetlands) are located on or adjacent to site, show detailed site restoration measures. Specify terrain, vegetation and trees (including size, spacing and species) which will be used to restore site to revegetated condition after foundation work and project completion.			
i.	**Stormwater/Utility Plan.** Utilities may be included on site plan or on a separate Stormwater/Utility Plan. Show stormwater systems, including drain lines, catch basins, watercourses, detention facilities, etc. Clearly draw all *existing* and *proposed* utilities such as side sewer, water service, fire hydrants, etc.			
j.	**Tree Plan.** Trees may be shown either on the site plan or on a separate Tree Plan. Must show the location, diameter and species of significant trees (conifers >6 feet tall or deciduous trees >6 inches in diameter at 4 ½ feet above the ground), including trees on site and in adjacent rights of way. Clearly designate all eagle perch/nest trees. Draw an "X" through trees to be removed and note tree protection fencing for trees near construction activities.			
k.	**Additional Details** as necessary with all details clearly referenced on the building plans and no notes or details on the plans that are not used for this project.			

S:\DSG\FORMS\SubmittalCOMM 01/2013

Figure 2.2 Submittal checklist for commercial projects.

Process for Submitting Plans for Commercial Project

Below are the 6 steps for submitting an application. Steps 1 and 2 are optional but highly recommended. Step 4 may or may not be required depending on the scope of your project. It is the goal of the Development Services Group (DSG) to provide you with as much information and assistance as possible in advance of your application submittal. This will ensure that the application contains information necessary to complete a thorough review in a timely manner, with minimal correction notices, requests for additional information, and other delays. Remember, all applicable land use applications must be approved before building plans may be submitted for review.

Step 1 – Review Site History

- Check your title report for special conditions or restrictions related to the development of your property including covenants, easements or other conditions.
- Ask to see the "Street File" for your project site at the front counter of DSG. The City has individual street files for most of the lots on Mercer Island with information on past development and activities at the site.
- Ask if there are any outstanding permits on the property.
- Look at the maps in DSG to find out more information about zoning, eagle habitats, utility locations, geologically hazardous areas, critical areas, stormwater drainage and streams.

Step 2 – Talk to Staff in DSG about Your Project

- Find out more about the process of submitting and reviewing plans with a Permit Coordinator, including specific code and application requirement, fee estimates, and estimated turnaround times.
- Talk with a Planner about any planning and zoning issues associated with your project. Ask any questions you may have regarding setbacks, lot coverage, gross floor area, average building elevation and maximum building height. *If this is a non-conforming structure or site or if a Variance or Deviation is required, you must complete the land use process with a Notice of Decision approval before submitting building plans for review.*
- Talk with a Development Engineer about utilities and earthwork on the site. Find out about existing utilities and utility work requirements for your project.
- Contact the permit center to discuss fire and safety requirements including access to your site, water flow, smoke detector/fire sprinkler requirements, and fire protection systems.
- Talk with the City Arborist if any trees will be removed as a result of your construction project. Find out what trees would need to be replanted and which trees need to be protected during the construction. Determine if your site contains any eagle perch trees or eagle nesting trees.
- Meet with the Building Plans Examiner to discuss your building plans, including structural calculations, increasing loads on existing foundations, energy calculation, geotechnical reports, and other building code issues.

Step 3 – Complete ALL of the Application Materials

- Pick up a packet of information and application materials and fill out the paperwork in advance of the pre-application meeting. Refer to the "Application Submittal Checklist" for a list of what materials are required for submittal.

S:\DSG\FORMS\SubmittalCOMM 01/2013

Figure 2.2 (*Continued*)

Step 4 – Schedule a Pre-Application Meeting

- Pre-application meetings are held once a week on Tuesdays, each 50-minute appointment is for a single site only.
- There is a $415.09 charge for each pre-application meeting and an additional $207.55 charge for each additional pre-app lication meeting required due to incomplete or ins ufficient application materials, missed appointments or cancellation with less than 24 hour prior notice.
- The pre-applic ation m eeting is designed to have submittal accomplished in one visit. Applicants are expected to arrive with complete s ubmittal materials. You will meet with various DSG staff including a Planner, Development Engineer, Building Plans Examiner and Permit Coordinator, who will review your project proposal.
- See the handout titled "Pre-Application Meeting Information" for more information about scheduling a pre-application meeting with the City of Mercer Island.

Step 5 – Submit Application and Plans

- After the pre-application meeting is finished, you may be able to submit your plans IF your submittal packet is complete, all applicable land use actions have been approved, and no additional information about the project is needed from DSG staff.
- If staff requires additional information or any changes to the plans, make the changes and then call your Permit Coordinat or to schedule a meeting to submit (once all the changes have been made).
- At the time of submittal, you will be required to pay the fees for plan checking. This is roughly one-third of the fee for the building permit, based on the valuation of the project. The remaining two-thirds will be due when the permit is issued.

Step 6 – Checking on Project Status

- After the permit is submitted, the Development Services staff will review the proposed project to ensure it meets all City regulat ions as well as current building and fire codes. The project may be reviewed by the Planner, Development Engineer, City Arborist and Building Plans Examiner, depending on the project's scope.
- You can c heck on the status of your permit by calling (206) 275-7605 and asking to speak with a Permit Coordinator. The normal turnaround time for a commercial project is 8-10 weeks. This time is estimated based on past projects. During the busier times of the year when many projects are being submitted (usually April through August), review times may be longer. Similarly, if y ou have an unusually complex project, submit in phases, or submit several corrections, the review time will also generally be longer.
- When your permit is ready to be picked up, a Permit Coordin ator will conta ct you. He/she can tell you if any other paperwork or information is required before the permi t can be issued and what fees will need to be paid at the time the permit is picked up.

S:\DSG\FORMS\SubmittalCOMM 01/2013

Figure 2.2 (*Continued*)

In accordance with Section 1803.5 of the IBC, geotechnical investigations might be required under the following conditions:

- Questionable soil
- Expansive soil
- High groundwater tables
- Deep foundations
- Rock strata
- Excavation near foundations
- Compacted fill material
- Controlled low-strength material (CLSM)
- Alternate setbacks and clearances near slopes
- High-seismic regions

Work involving existing buildings could often meet one or more of the conditions listed above and would therefore require that a geotechnical investigation be performed. As an example, see Fig. 2.3 which depicts a recent project that was completed on one of the campuses for Salt Lake Community College (SLCC) in the state of Utah. The project called for a significant alteration and several additions to an existing unreinforced masonry building. Several of the additions were built adjacent to existing construction and required excavations near existing foundations. In addition, there were several new shear walls constructed within the existing space that were supported by new foundations. Micropiles were used to support

Figure 2.3 SLCC Center for New Media. (*Big D Construction, n.d.*)

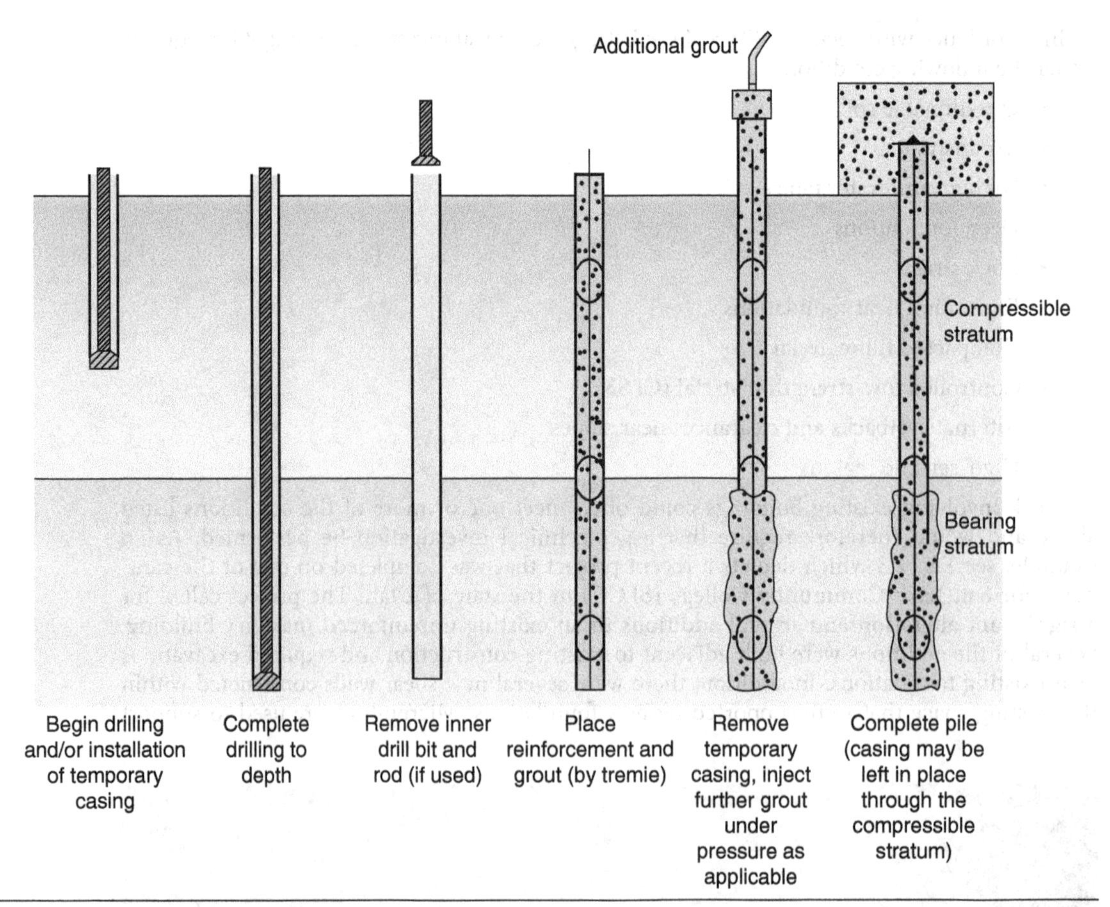
Additional grout
Compressible stratum
Bearing stratum
Begin drilling and/or installation of temporary casing
Complete drilling to depth
Remove inner drill bit and rod (if used)
Place reinforcement and grout (by tremie)
Remove temporary casing, inject further grout under pressure as applicable
Complete pile (casing may be left in place through the compressible stratum)

Figure 2.4 Micropile installation process.

the new foundations and to underpin the existing foundations in some instances. As defined within the IBC, micropiles are deep foundation elements that have a diameter of no more than 12 in. and consist of high-strength grout and central steel reinforcement. Figure 2.4 displays the typical micropile installation process as portrayed in the Federal Highway Administration's (FHWA) Micropile Design and Construction reference manual (FHWA, 2005). For the SLCC Center for New Media project, a detailed geotechnical investigation was required to allow both the structural engineer and the micropile designer to complete their portion of the design.

In addition to the requirements outlined in Section 1803.5 of the IBC, other referenced standards may require that a geotechnical engineer be involved on a project. As an example, Section 303 of the IEBC allows seismic evaluations to be performed in accordance with either the IBC or with ASCE 41-17, Seismic Evaluation and Retrofit of Existing Buildings, published by the American Society of Civil Engineers. There are several instances within the ASCE 41-17 standard where a geotechnical investigation is required. Section 8.2.2 of ASCE 41-17 requires an "...in situ geotechnical investigation shall be performed..." anytime there could be earthquake-induced hazards at a site (ASCE, 2013, 124). These hazards could include surface fault rupture, liquefaction, differential settlement, landslides, or even the potential for inundation during a tsunami event.

2.3.3 Structural calculations. When repairs, alterations, additions, or changes in use affect the structural members of the building, structural calculations should be provided. These calculations could be to simply show that the existing structural elements are adequate or can be for the installation of new structural members. Calculations are almost always required when moving an existing structure in accordance with Chapter 14 of the IEBC. All throughout the IEBC, it notes when a structural evaluation or analysis is required. Anytime this requirement is triggered, structural calculations must be provided with the permit application.

Many times, the code official may not be comfortable reviewing the structural calculations, but regardless they should ensure that these are submitted in such a fashion that shows that an adequate analysis has been performed. In most states, the professional licensing statutes require that calculations be sealed when submitted as part of a building permit application. An engineer's seal typically comprises a professional engineering stamp that lists the individual's license number, professional discipline (e.g., civil, structural, mechanical, electrical, etc.), and the state in which they are registered, and then their signature and date are typically placed across the seal.

2.3.4 Energy compliance reports. In most cases the IEBC requires that all new construction that is performed shall comply with the requirements of the IBC. The only exception to this is for repairs that are considered "less than substantial," which we will discuss in more detail later in this text. It is further noted within the IEBC itself that all alterations and additions within existing buildings must be shown to comply with the energy requirements outlined in the IECC, or the IRC for single-family homes. Chapter 5 of the IECC has specific energy provisions for existing buildings.

Generally, Chapter 5 of the IECC requires all new or altered portions of existing buildings to meet the requirements for new buildings. In the case of alterations, there are some exceptions in which, based on the scope of the alteration, existing elements are not required to be updated. For additions the designer must show that either the addition complies with the requirements for new construction or that the altered building as a whole complies. The most common way to show compliance with the energy provisions is to meet certain mandatory requirements in addition to providing an energy compliance report that shows the building complies with the prescriptive requirements of the IECC or the provisions of ASHRAE 90.1. This alternate standard is published by the American Society of Heating, Refrigerating, and Air-Conditioning Engineers (ASHRAE) and outlines energy requirements for commercial buildings.

While small alterations or repairs to an existing building will likely show compliance with the energy provisions by including certain prescriptive elements outlined in Chapter 5 of the IECC directly on the plans, significant alterations or additions will likely need to include energy compliance reports in addition to the plans in order to show compliance. The most common software used is entitled COMcheck™, developed by the U.S. Department of Energy. This software can demonstrate compliance with the provisions of the IECC or ASHRAE 90.1.

2.3.5 Special inspection and structural observation programs. There is a significant difference between Section 106.1 of the IEBC and the same section in the IBC. In the IBC it states that submittal documents should include "...construction documents, *statement of special inspections*, geotechnical report, and other data," while the IEBC states that it should include "...construction documents, *special inspection and structural observation programs, investigation and evaluation reports*, and other data." While it may sound similar, there is a significant distinction between what should be submitted for projects involving existing buildings versus new buildings. For now, let's focus on the differences in relation to special inspections and structural observations.

While Chapter 1 of the IEBC makes reference to special inspections being required for existing buildings, the special inspection requirements are not found within the IEBC. To determine the special inspection requirements for a project, the user must depend upon what is included in the IBC. The following will highlight the IBC requirements, while focusing on what items may need to be considered in relation to existing buildings.

Section 202 of the IBC defines special inspections as the "Inspection of construction requiring the expertise of an approved special inspector." These are inspections that are performed by a private third-party entity that is hired by the project owner. Sections 1703.1 and 1704.2.1

of the IBC require the building official to approve special inspection agencies and the actual special inspectors that work for them. Approval is typically based upon the inspector's work experience, as well as their possession of a certification that shows they are qualified to oversee specific portions of the work.

Figure 2.5 shows an image of a concrete special inspector and provides a good example of the specialized experience required to serve as a special inspector. To serve as a concrete special inspector, the individual must take classes and pass a test to become certified by a national organization such as ICC or the American Concrete Institute (ACI). In this picture the special inspector is currently making concrete cylinders that will later be broken in a laboratory to ensure that they reach the required strength as specified on the project plans. He is also checking the air content in the concrete using the item in front of him that looks like a pot and has a gauge affixed to the top. At the bottom of the picture is a plate that was used as a base to check the "slump," or how fluid the concrete is, as it leaves the back of the mix truck. All these are important constructability items and could require field modifications if they do not meet the tolerances specified in the approved construction documents.

Figure 2.5 Concrete special inspections. (*Basham, 2013.*)

Section 1704.3.1 of the IBC requires that all projects requiring special inspections include a Statement of Special Inspections (SSI). This means that all items requiring special inspections must be listed, that the extent of those inspections be identified, and that the frequency of the required inspections be noted. One should not state that only concrete special inspections are required. Figure 2.6 displays Table 1705.3 of the IBC. This table highlights the extent of the

Required Special Inspections and Tests of Concrete Construction				
TYPE	CONTINUOUS SPECIAL INSPECTION	PERIODIC SPECIAL INSPECTION	REFERENCED STANDARD[a]	IBC REFERENCE
1. Inspect reinforcement, including prestressing tendons, and verify placement.	—	X	ACI 318: Ch. 20, 25.2, 25.3, 26.6.1-26.6.3	1908.4
2. Reinforcing bar welding: a. Verify weldability of reinforcing bars other than ASTM A706; b. Inspect single-pass fillet welds, maximum 5/16"; and c. Inspect all other welds.	— X	X X	AWS D1.4 ACI 318: 26.6.4	—
3. Inspect anchors cast in concrete.	—	X	ACI 318: 17.8.2	—
4. Inspect anchors post-installed in hardened concrete members.[b] a. Adhesive anchors installed in horizontally or upwardly inclined orientations to resist sustained tension loads. b. Mechanical anchors and adhesive anchors not defined in 4.a.	X	 X	ACI 318: 17.8.2.4 ACI 318: 17.8.2	—
5. Verify use of required design mix.	—	X	ACI 318: Ch. 19, 26.4.3, 26.4.4	1904.1, 1904.2, 1908.2, 1908.3
6. Prior to concrete placement, fabricate specimens for strength tests, perform slump and air content tests, and determine the temperature of the concrete.	X	—	ASTM C172 ASTM C31 ACI 318: 26.5, 26.12	1908.10
7. Inspect concrete and shotcrete placement for proper application techniques.	X	—	ACI 318: 26.5	1908.6, 1908.7, 1908.8
8. Verify maintenance of specified curing temperature and techniques.	—	X	ACI 318: 26.5.3-26.5.5	1908.9
9. Inspect prestressed concrete for: a. Application of prestressing forces; and b. Grouting of bonded prestressing tendons.	X X	— —	ACI 318: 26.10	—
10. Inspect erection of precast concrete members.	—	X	ACI 318: 26.9	—
11. Verify in-situ concrete strength, prior to stressing of tendons in post-tensioned concrete and prior to removal of shores and forms from beams and structural slabs.	—	X	ACI 318: 26.11.2	—
12. Inspect formwork for shape, location and dimensions of the concrete member being formed.	—	X	ACI 318: 26.11.1.2(b)	—

For SI: 1 inch = 25.4 mm.

a. Where applicable, see Section 1705.12, Special inspections for seismic resistance.

b. Specific requirements for special inspection shall be included in the research report for the anchor issued by an approved source in accordance with 17.8.2 in ACI 318, or other qualification procedures. Where specific requirements are not provided, special inspection requirements shall be specified by the registered design professional and shall be approved by the building official prior to the commencement of the work.

Figure 2.6 Table 1705.3 of the IBC. (*Excerpted from the 2018 International Building Code; copyright 2017. Washington, D.C.: International Code Council.*)

concrete special inspections that could be required on a project and notes the frequency of the associated inspections. A true project SSI should include something similar to this table for all items requiring special inspections.

It is important that the user understand that Chapter 17 of the IBC is not an all-encompassing document that lists all items that may require special inspections. Rather, Section 1705.1.1 of the IBC is provided to highlight the fact that there may be several items that may require special inspections that are not included in the chapter. When dealing with existing buildings, there will be items that one does not come across every day that may require special inspections. The National Council of Structural Engineers Associations (NCSEA) currently has a committee that is looking into adding specific special inspection items to the IEBC. For now, the following are just a few examples of items that may require special inspections in relation to existing buildings:

- *Moved Buildings:* Special inspection of structural components and connections to ensure that structural damage was not sustained during the moving operations.
- *Unreinforced Masonry Bearing Walls:* Special inspection for new anchors in unreinforced masonry as well as mortar repointing operations. Unreinforced masonry buildings are typically those that were constructed prior to 1975, as subsequent codes have required additional reinforcement and detailing in masonry construction. Figure 2.7 shows the repointing operations that were performed at the Greek Orthodox Church of St. Nicholas in Ballarat, Australia. Repointing is also known as "tuck pointing" and requires the removal

Figure 2.7 Brick mortar repointing. (*Basham, 2013.*)

and replacement of deteriorating mortar. Not only does repointing improve the building's appearance, but it also significantly improves the shear capacity of the masonry wall.

- *Adjoining Buildings:* New York City has special requirements for adjoining buildings, as they have had numerous incidents when construction occurs to existing buildings that adjoin (Eschenasy et al., 2016). They highlight that third-party special inspections should occur if existing foundations will be undermined, if new forces will be applied to existing elements, or if any movement (i.e., vibrations) will be introduced to adjoining properties as part of the work. Section 1704.20.1 was added to the *New York City Building Code* to address this condition and the structural stability of existing buildings in general.
- *Adjacent Properties:* Monitoring should be provided of existing structures that are adjacent to construction operations. An adjacent structure could include a building, utilities, roads or sidewalks, or even buried structures. Common sources of damage to adjacent structures include excavations; dewatering activities; and vibrations due to demolition, blasting, pile driving, or drilling operations. A monitoring program should be developed during construction that would check for vibration and movement. The plan should indicate the types and locations of monitoring equipment, the frequency of the readings, threshold criteria, and an action plan if criteria are exceeded (Smith, 2015). Figure 2.8 displays the deep foundation excavations that were performed at the New Science Building for George Washington University. In the image one can see the adjacent brick buildings that would likely require monitoring during the construction activities of the new building.

Figure 2.8 New Science Building at George Washington University. (*ForConstructionPros.com, 2015.*)

- *Fiber-Reinforced Polymers (FRP):* When used to strengthen existing concrete or masonry structural elements, appropriate special inspections should be provided during installation of fiber-reinforced polymers. For example, Fyfe Company has a product called Tyfo® Fibrwrap®, and they have gone through the process of vetting their product with ICC's Evaluation Services. As such, they have an evaluation report (ESR-2103) that clarifies that special inspections are to be provided during installation and outlines the specific special inspection procedures.

As noted previously, Section 106.1 of the IEBC calls for both special inspections and structural observations. Section 202 of the IBC defines structural observations as "The visual observation of the structural system by the (Structural Engineer of Record – SER) for general conformance to the approved construction documents." Perhaps the most important aspect of quality assurance in relation to existing buildings is to ensure that the SER is onsite to perform structural observations as often as possible. Structural observations are discussed in Section 1704.6 of the IBC, but this is typically limited to rather large buildings unless specified by the SER or the code official.

Like Section 106.1 of the IEBC, Section 1.5.10.1 of ASCE 41-13 states that the SER should note the structural observation requirements on the plans, and Section 1.5.10.2 of ASCE 41-13 goes on to say that the SER should provide periodic structural observations (ASCE, 2013, 28–29). For new construction, these observations are mainly to ensure that construction is in conformance with the approved construction documents, while with existing construction this takes on the added requirement of confirming that the existing conditions match what was assumed in the structural design.

In accordance with Section 1704.6 of the IBC, the SER is required to provide structural observation reports to the code official noting any items that they feel are deficient and have not been resolved. Figure 2.9 is an example of what the structural observation report could look like. This sample report was developed by the Clark County, Nevada Department of Building and Fire Prevention (Clark County, 2017).

2.3.6 Investigation and evaluation reports. In addition to special inspections and structural observations, Section 106.1 of the IEBC states that investigation and evaluation reports may be part of the submittal documents for existing buildings. While much more detail will be provided later in this book, by means of the IEBC, the code official could require either an investigation or evaluation report for any of the following purposes:

- *Overall Building Evaluation (IEBC § 104.2.2.1):* The purpose of this would be to identify any items that are not in conformance with the IEBC.
- *Seismic Evaluation (IEBC § 303):* There are several instances within the IEBC when it states that a seismic evaluation is required. Section 303 outlines the procedures that should be followed as part of this evaluation.
- *Damaged Buildings (IEBC § 405.2.3.1):* This evaluation requires a licensed professional engineer to review a damaged building and to determine if it can be repaired to its pre-damaged state or if additional upgrades would be required.
- *Unreinforced Masonry (IEBC § 503, 706.3.1, 906):* There are triggers throughout the IEBC that would require specific aspects of an unreinforced masonry building to be evaluated. The required evaluations could include whether bracing is required to parapet walls or if new wall connections are required at the roof and floor lines. Figure 2.10 includes an image that was published in the *Seattle Times*, August 10, 2015 (Doughton and Raghavendran, 2015). This image highlights key wind and seismic strengthening measures that are often recommended in relation to existing unreinforced masonry buildings.

DEPARTMENT OF BUILDING & FIRE PREVENTION
STRUCTURAL OBSERVATION REPORT

PERMIT NO:________________ ADDRESS: ____________________

OBSERVER FIRM NAME:________________ OBSERVER NAME: ________________

This form was submitted to the following individuals.

Entity	Name	Entity	Name
CCDB		Owner's Representative	
Contractor		Special Inspector	

☐ **First Observation** (Date) ________________

☐ **Discrepancies Observed** (Date) ________________

(List discrepancies) ________________

☐ **Discrepancies Resolved** – Resolution(s) are in accordance with the approved construction documents (list resolution and date).

☐ Structural Observation Approved

I hereby acknowledge that I performed the structural observations in accordance with the approved structural observation plan and TG-10. The work that has been observed complies with the intent of the approved construction documents.

Observations (describe the observations and locations; foundation, level 1, gridlines) ________________

Attachments ________________

REGISTERED DESIGN PROFESSIONAL SEAL HERE

Form 802 7/12/2016

Figure 2.9 Structural observation report—Clark County, Nevada.

Retrofitting old brick buildings

Upgrading an old building to make it more earthquake-safe involves connecting brick walls and parapets to the roof and floors.

REQUIRED

❶ **Parapet bracing:** The portions of a wall that extend past the roof (parapet) need a diagonal bracing that is generally made of steel.

❷ **Wall-to-roof diaphragm anchors:** Steel bolts horizontally secure the brick wall to the roof. Rosettes seen on the outside of a building can indicate that this retrofit has been done.

❸ **Wall-to-floor diaphragm anchors:** Steel bolts tie the brick wall to the floors.

SOMETIMES REQUIRED

❹ **Out-of-plane wall bracing:** Steel beams that vertically connect the brick wall to the floors to keep the wall from bending.

❺ **Overall building bracing*:** Steel beams that increase a building's overall strength.

❻ **Diaphragm strengthening*:** Plywood sheathing that strengthens floors and roofs.

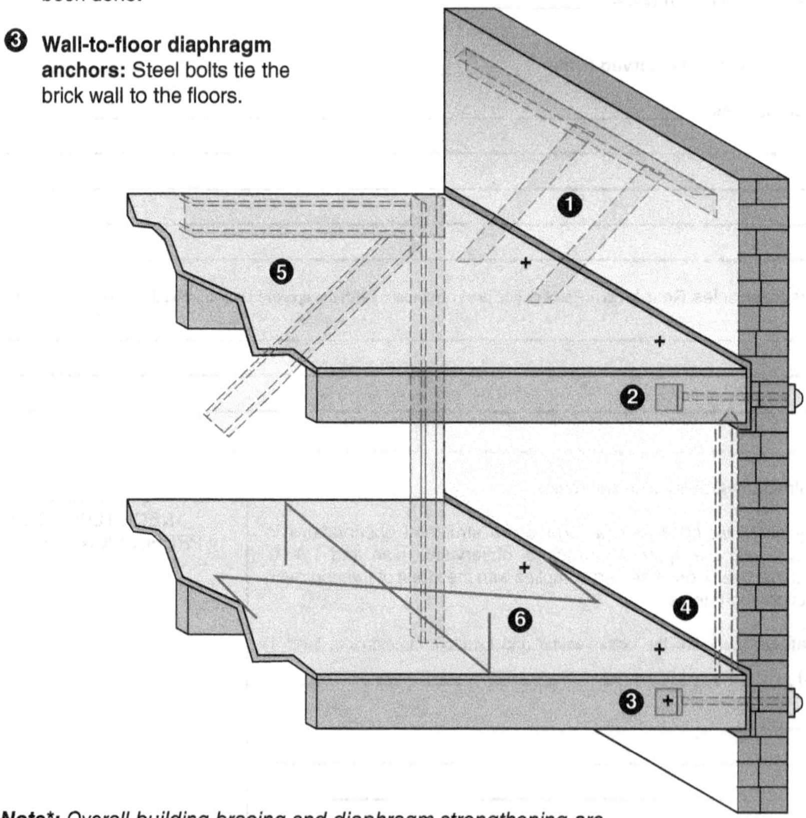

***Note*:** *Overall building bracing and diaphragm strengthening are often important for buildings with large windowed shops on the first level.*

Figure 2.10 Unreinforced masonry retrofit elements.

- *Reinforced Concrete or Masonry Walls (IEBC § 503.7, 906.4):* In some instances, the IEBC requires that buildings with concrete or masonry walls be evaluated to determine whether additional wall anchors are required along the roof line. This trigger only occurs in high-seismic regions but could be required on much newer buildings than the requirements for unreinforced masonry buildings.
- *Roof Diaphragms (IEBC § 503.12, 706.3.2):* When removing roofing materials from at least 50 percent of the roof, the IEBC may require that an evaluation be performed showing that the existing roof framing members, roof-to-wall connections, and wind uplift anchorage is adequate. This requirement is only for special wind regions or areas where the design wind speed is greater than 115 mi/h.
- *Fire-Resistance Ratings (IEBC § 802.6):* When an applicant chooses to apply the current IBC fire-resistance ratings to an existing building, an investigation and evaluation report must be provided to the code official noting any special construction features that could impact said fire-resistance ratings.
- *Historic Buildings (IEBC § 1201.6):* If required by the code official, an overall evaluation of the historic building must be provided that identifies items that are not in compliance with the code and note, that if corrected, if they would be "...damaging to the contributing historic features."
- *Performance Compliance Method (IEBC Chapter 13):* When following the performance path to conformance, the building owner is required to perform a detailed investigation of the existing structure and to provide an overall evaluation that includes an analysis of the following disciplines: structural, fire safety, means of egress, and general safety.

2.4 *Inspections*

Like Section 110 of the IBC, Section 109 of the IEBC outlines when code inspections are required. Code inspections refer to the inspections that are performed by the local building authority and do not include third-party special inspections discussed earlier.

The permit holder must contact the building department to schedule an inspection when the work is ready for inspection. They must provide the code inspector access to the items requiring inspection, and they must remain visible and cannot be covered up until the work has been approved. Many jurisdictions require more than 24-h notice to schedule an inspection.

The following inspections are specifically required by the IEBC:

- *Footing and foundations (IEBC § 109.3.1):* The code inspector should be notified by the contractor to perform an inspection only after all footings have been excavated, formwork has been placed, and reinforcing steel is properly placed and tied. As part of this inspection, the code inspector will verify that foundation elements are of the proper size and in the correct locations, that reinforcement steel is installed correctly, and that the general installation is in accordance with the approved plans and complies with the code. No concrete should be poured until after the footing and foundation inspection has been approved in writing by the code inspector.
- *Concrete Slab or Underfloor (IEBC § 109.3.2):* Most concrete slabs contain reinforcing steel, conduits, piping, or other items within the slab. Prior to placing concrete, it is important to ensure that these items are properly installed and tied in place. When dealing with existing buildings, it is quite common that new underground plumbing systems will be installed (see Fig. 2.11) and these systems must be checked prior to placing concrete. If a vapor barrier is required by the local jurisdiction, the installation of such should also be inspected at this time.

Figure 2.11 **Underslab plumbing.**

- *Lowest Floor Elevation (IEBC § 109.3.3):* This inspection is only required where construction is to occur within an area that could be subject to flooding. When this occurs, the code inspector must verify that the finished floor level will be above the Base Flood Elevation (BFE) as shown in Fig. 2.12. The BFE signifies a hypothetical elevation to which flood waters might rise during a 100-year flood event. These elevations are established by the Federal Emergency Management Agency (FEMA) on regional maps called Flood Insurance Rate Maps (FIRM). For existing buildings, the lowest floor elevation

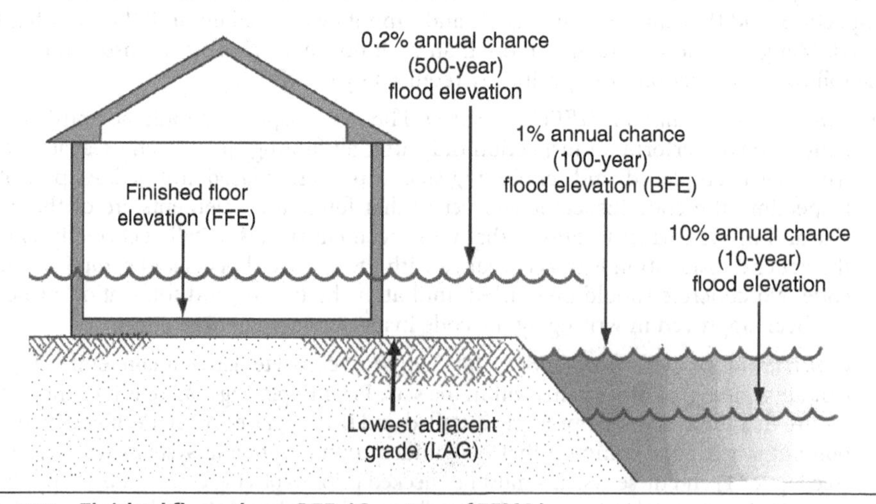

Figure 2.12 **Finished floor above BFE. (*Courtesy of FEMA.*)**

inspection would only be required for alterations or additions that constitute a substantial improvement (i.e. 50 percent of the structure's value or more).

- *Frame Inspection (IEBC § 109.3.4):* During the frame inspection, the code inspector ensures that the structural framing members are installed correctly. As part of this inspection, they ensure that members are of the appropriate size and that the correct fasteners and connections have been used as outlined on the approved construction documents. This inspection is typically performed after all plumbing, mechanical, and electrical systems have been installed so that the inspector can ensure that those trades did not adversely affect the structural elements before they are covered.
- *Lath or Gypsum Board (IEBC § 109.3.5):* When gypsum board is installed as part of a shear wall, fire-rated wall, or fire-rated horizontal assembly, the code inspector must inspect the installation before any joints or fasteners have been taped or finished. For fire-rated installations, the inspector should review the listings to ensure that the correct type of gypsum product is used and the entire fire-rated assembly has been installed correctly.
- *Weather-Exposed Balcony (IEBC § 109.3.6):* This inspection requirement is new to both the 2018 IBC and the 2018 IEBC. This requirement was brought about after a balcony collapsed on June 16, 2015, in Berkeley, California, killing six young people and injuring seven others. Figure 2.13 is an image of the failed balcony in Berkeley. The investigation that was performed after the collapse found that improper materials were used and that the balcony was not properly waterproofed (Orenstein, 2017).

Figure 2.13 Berkeley, California, balcony collapse.

This new code section requires that the code inspector verify that impervious moisture barrier systems are properly installed, that positive drainage exists, and that cross ventilation is provided. This requirement does not apply to balconies or decks that use standard decking materials that allow the balcony or deck to be self-draining. Section 109.3.6 of the IEBC also highlights that the code official may require this inspection to be performed by a special inspector in accordance with Section 1705.1.1 of the IBC.

- *Fire- and Smoke-Resistant Penetrations (IEBC § 109.3.7):* Any holes that are created within a fire-rated assembly are considered a penetration. It is important that the code inspector take a significant amount of time to ensure that all openings within a fire-rated assembly, whether they be walls or floors, are properly sealed. Figure 2.14 shows the annular space around piping and the joint between the floor and the wall being properly sealed. To properly seal the opening, the contractor must use approved materials that allow the assembly to maintain its fire-resistance rating. As an example, if a pipe penetrates a 2-h fire wall, a listed 2-h rated fire caulk will be required to seal the annular space.
- *Other Inspections (IEBC § 109.3.8):* In addition to the inspections listed above, the code official is authorized to require other inspections that may be needed to ensure that construction is in compliance with state and local requirements. All jurisdictions, however, should clearly share with the permit applicants what inspections are required ahead of issuing a building permit. The *Connecticut Building Code* has been amended to require a schedule of all required inspections to be compiled by the code official and posted in the building department for public view (Currey, 2016).

Figure 2.14 Fire-resistant penetrations. (*Firestop, n.d.*)

The following is a partial list of other inspections that are often required by local jurisdictions:

- Foundation drain and waterproofing inspection
- Insulation and vapor barrier inspection
- Suspended ceiling inspection
- Rough-in inspection

It should be noted that while the IBC requires energy efficiency inspections, the IEBC does not. Similar to the IBC, however, the IEBC does require that a final inspection be performed. This inspection must be performed prior to occupancy of the building and often requires multiple visits by the code inspector. The purpose of the final inspection is to confirm that all work in relation to the building permit has been completed in accordance with the code and approved plans.

2.5 *Certificate of Occupancy*

For new buildings, a certificate of occupancy is always required prior to occupying the building. In relation to existing buildings, a new certificate of occupancy must be issued for altered areas of the building, relocated buildings, or buildings that undergo a change in use.

The code official should not issue a certificate of occupancy until all items in relation to the building permit have been completed. This often requires a final code inspection report, a final special inspection report, a certificate of fire clearance, sign-off on all deferred submittal items, and sign-off from other authorities such as the Health Department or State Elevator inspector.

Figure 2.15 displays a sample certificate of occupancy for a hypothetical jurisdiction in the state of Ohio. In accordance with Section 110.2 of the IEBC, the following items must be included on the certificate of occupancy that is issued, while the local jurisdiction may choose to note additional items as well:

- The building permit number
- The address of the structure
- The name and address of the owner
- A description of the portion of the structure for which the certificate is being issued
- A statement that it has been inspected for code compliance in relation to the proposed occupancy type
- The name of the code official
- The edition of the code for which the permit was issued
- The use and occupancy of the building
- The type of construction
- The design occupant load
- Listing of fire protection systems installed
- Any special stipulations or conditions

Certificate of Occupancy

Office of the Building Official

Property Address:

Approved As:
Pre-Existing Condition (No Change)
New Structure
Alteration
Change of Occupancy
Temporary Occupancy

Use Groups:	**Occupancy Description:**
Primary:	
Accessory:	
Accessory:	
Mixed Uses:	

Attached Floor Plan dated__________indicates of how areas are approved and design occupancy loads

Construction Type:

Fire Sprinkler Systems:

N/A ☐	Required ☐	Non-Required ☐
System Type:		Location:

Hazard Classification:

Storage Height:

Aisle Width:

Sprinkler System Demand @ base of riser:

Standpipe System Demand @ base of riser:

Stipulations, Conditions, Variances:

This Certificate represents an approval that is valid only when the building and its facilities are used as stated and is conditional upon all building systems being maintained and tested in accordance with the applicable Board of Building Standards rules and applicable equipment or system schedules.
This certifies conformance with Chapters 3781. and 3791. of the Revised Code and the applicable provisions of the rules of the Board of Building Standards.

Plan Approval Application #__________
Approved pursuant to the following editions of:
__________ OBC __________ OMC __________ OPC

This approval is limited to the following portion of the building:

The balance of the building is approved pursuant to the following dated C of Os:

Date:

Building Official:

Figure 2.15 **Example certificate of occupancy.**

2.6 *Board of Appeals*

The building codes are not always black and white. There are several instances where the reader can interpret a specific code requirement one way, while another could interpret it very differently. With this in mind, it is understandable that there will be times that the permit applicant does not agree with an interpretation made by the code official. For this purpose, Section 112 of the IEBC allows for the establishment of a building board of appeals. Appendix B of the IBC provides a great deal of information as to how a jurisdiction can set up a board of appeals and addresses items such as qualifications, rules and procedures, how to issue notices, and how decisions are made.

The building board of appeals should be composed of persons who are not employed by the local jurisdiction and who are qualified by means of experience, training, or education. Most building boards are made up of licensed design professionals, contractors, or other code officials. The purpose of the board is simply to determine if the code has been correctly interpreted by the code official. The board has no authority to waive a code requirement.

The code does not specify a minimum number of board members, nor does it lay out any specific governing rules. Each jurisdiction should establish their own building board of appeals and rules governing the board. Section 125.1514 of the Construction Code Act for the state of

Michigan establishes the minimum requirements for boards of appeals in Michigan (State of Michigan, 1972). This statute requires that the board consist of between three and seven members that are appointed for 2-year terms. It also established how one would submit an application for appeal and that the appeals hearing should take place within 30 days of the application.

2.7 *Violations and Stop Work Orders*

There are many instances when a code official may need to issue a violation, or a stop work order, on a project involving an existing building. Section 113 of the IEBC notes that a violation occurs anytime work is being performed that is contrary to a particular code requirement. The code official should issue a notice of such violations, and penalties could occur if they are not remedied in a prompt manner. A stop work order is issued by the code official anytime the work being performed is deemed to be dangerous or unsafe. This shall be delivered in writing, and the work in question should immediately cease.

2.8 *Unsafe Buildings*

Sections 115 through 117 of the IEBC outline provisions where the code official is authorized to declare a building unsafe and to require either emergency measures or demolition to occur. Section 115.1 states that unsafe buildings "...shall be taken down, removed, or made safe." Obviously, such a stance should not be taken lightly by the code official and likely requires authorization from others within the jurisdiction before proceeding.

For the code official to declare a building unsafe, they must serve an official notice to the owner, or the owner's authorized agent. The notice must identify the items that are unsafe and the corrections that must be made. The notice should also specify a time frame in which the repairs must be completed before demolition might be required. Once in receipt of the notice, the owner must contact the code official and state their acceptance or rejection of the terms.

A code official would never arbitrarily declare a building unsafe. Before such a step is taken, they would be meticulous and would document carefully how the unsafe condition has been triggered. A building is typically labeled unsafe when it is unsanitary or unfit for human occupancy. In some instances, the city will placard the building and put up temporary fencing noting that it is unsafe as shown in Fig. 2.16. The unsafe triggers are outlined in the definitions of "unsafe" and "dangerous" in Section 202 of the IEBC. The following are common triggers for declaring a building or structure unsafe:

- Imminent danger of failure or collapse
- Any structure, or portion thereof, that has collapsed, partially collapsed, moved off its foundation, or lacks support of the ground
- Unsanitary
- Inadequate means of egress
- Inadequate light or ventilation
- Constitutes a fire hazard
- Structure dangerous
- Attractive nuisance

Figure 2.16 Unsafe building sign. (*BBC, 2010.*)

An article written in *The News Tribune* on March 4, 2018, noted that Jefferson City, Missouri, declared about one to two buildings unsafe per month (Roberts, 2018). Most of these are brought to the code official's attention by the police or fire departments. While one to two a month might sound like a lot, the same article mentioned that more than 3200 code violations were issued during the 2017 fiscal year and that the city does everything it can to work with the building owner to get issues resolved prior to declaring it unsafe. The code official, Dave Helmick, was quoted as saying, "It's just 100 percent caring about the safety of the community, it's not just the owner or occupant of that house. It's the neighbors, it's the kids playing in the backyard."

CHAPTER

3

COMMON PROVISIONS

3.1 *Introduction*

As shown in Fig. 3.1, the *International Existing Building Code®* (IEBC®) allows the user to select from three different paths of compliance. Any of these three paths can be used when an existing structure is undergoing an alteration, addition, or change of occupancy. In addition, both the prescriptive and work area compliance methods have specific provisions that can be used when dealing with historic structures, while the performance compliance method does not.

In the past, the prescriptive and work area compliance methods also included provisions for repairs and relocated buildings, but that has changed with the 2018 edition of the IEBC. If the existing building is undergoing repairs, it must comply with Chapter 4 of the IEBC, whereas relocated or moved buildings must comply with the requirements of Chapter 14. As such, a compliance path does not need to be selected for repairs or relocated buildings unless they are also undergoing alterations, additions, or changes in occupancy.

When selecting one of the three paths shown in Fig. 3.1, it is important to note that only one path can be selected. The structural engineer of record (SER) cannot choose to follow the work area compliance method while the architect chooses to comply with the nonstructural provisions of the prescriptive compliance method. All disciplines must follow the same path and should therefore work together to decide which is most appropriate for the project in question.

Rather than following one of the three compliance paths, Section 301.3 of the IEBC includes an exception that allows the code official to accept alterations to an existing building that comply with the laws that were in existence when the building was originally constructed. This exception is allowed only under the following circumstances:

- When allowed by the code official. Most code officials would likely not allow this exception to be used unless substantial information was provided.
- New structural members must comply with the material and construction requirements of the current *International Building Code®* (IBC®).
- Alterations to existing buildings that constitute a "substantial improvement," as defined in Chap. 1 of this book, and which are in flood hazard areas must comply with Section 1612 of the IBC or Section R322 of the *International Residential Code®* (IRC®).

The exception cannot be used when in lieu of meeting the structural provisions of the prescriptive compliance method or the structural requirements for alterations under the work area method.

At the risk of overemphasizing this exception, consider a single-family residence that was constructed in 1985 and was built under the 1982 *Uniform Building Code* (UBC). The home is in a high-seismic region, and the owner wishes to remove a large portion of an interior bearing wall to make things more open in their home. While not much in relation to alterations

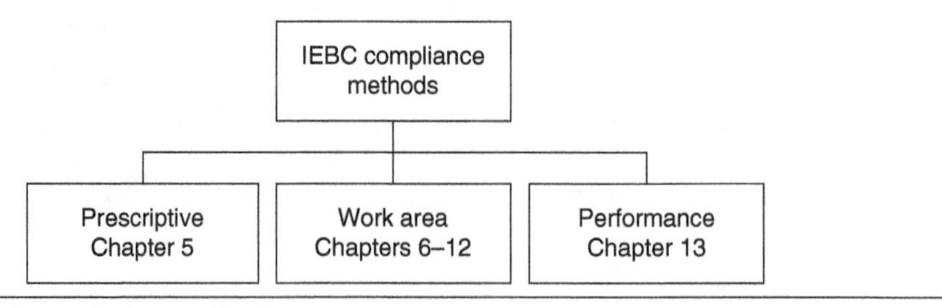

Figure 3.1 IEBC compliance methods.

is included in the 1982 UBC, Section 2312.j.2 allows minor alterations to be made but states that the lateral resistance cannot be reduced. If the lateral resistance is reduced as a result of the alteration, the entire structure would need to show compliance with the lateral requirements outlined in Section 2312 of the 1982 UBC. In this case, the owner verified that the interior bearing wall only supported gravity loads, so removing a portion of it would not affect the lateral resistance of the home. With this in mind, the new beam, support posts, and footings will need to be sized in accordance with the current IBC as noted in the second bullet above.

Considering Section 104.11, Alternative materials, design, and methods of construction, and equipment, and the "exception" to Section 301.3, it can be seen that in reality the IEBC provides flexibility for construction work in existing buildings through five options that can be chosen by the owner and the owner's design team. These include:

- Section 104.11
- Exception to Section 301.3
- Chapter 5: Prescriptive Compliance Method
- Chapters 6 through 12: Work Area Method
- Chapter 13: Performance Compliance Method

3.2 *General Provisions*

Section 302 of the IEBC provides general provisions that should be applied to all three compliance methods in addition to repairs and relocated structures. In previous versions of the IEBC, these items were often repeated throughout the code. It is important that the reader understands that each of the following items will always be required when performing any work under the IEBC.

3.2.1 Dangerous conditions. Section 302.2 of the IEBC states that the code official can require any dangerous condition to be eliminated. This section reiterates what is discussed in Sections 115 through 117 of the administrative provisions in the IEBC. Section 202 of the IEBC defines dangerous conditions as either a structure, or portion of a structure, that has collapsed, has partially collapsed, has moved off its foundation, lacks support, or where a significant risk of collapse exists.

As an example, Fig. 3.2 shows a partial building collapse that occurred in Alexandria, Virginia, on October 2, 2016. The building in question is a nine-story T-shaped condominium tower. One segment of the tower dropped 2 to 3 in. both vertically and laterally. The code official was notified and assisted other emergency personnel to evaluate the cause of the problem. They quickly found that the columns had buckled near the base due to years of water damage (Augenstein, 2016; Kilsheimer, 2017).

The residents of the 32 condo units were evacuated and displaced for several months. The code official worked closely with the condo association and the design team to ensure that all necessary repairs were made prior to allowing residents to return to their units.

As illustrated in this example, it is often a good idea for the code official to evaluate the building before any work occurs. While the owner of a building may have a plan for how to repair or improve an existing building, there might be other items that could be deemed "dangerous" that are not being addressed. While it is still the responsibility of the owner of the building and the design professional to evaluate and remediate such conditions, the code official should be actively engaged in the process.

Figure 3.2 Partial building collapse. (*Covering, 2016.*)

3.2.2 Additional codes. Section 302.3 of the IEBC clarifies that repairs, alterations, additions, changes of use, and relocated buildings will also need to comply with provisions included in other codes and not simply what is included in the IEBC. The following is a listing of the specific codes referenced within this section:

- *International Energy Conservation Code®* (IECC®)
- *International Fire Code®* (IFC®)
- *International Fuel Gas Code®* (IFGC®)
- *International Mechanical Code®* (IMC®)
- *International Plumbing Code®* (IPC®)
- *International Private Sewage Disposal Code®* (IPSDC®)
- *International Property Maintenance Code®* (IPMC®)
- *International Residential Code®* (IRC®)
- NFPA 70®—*National Electrical Code®* (NEC®)

Note that the IBC is not included in the list of additional codes referenced. As discussed in Chap. 1 of this book, the IEBC is a referenced code of the IBC (see Section 101.4 of the IBC). As such, the IEBC is truly an extension of the IBC. Section 102.4 of the IBC states that the referenced codes should be considered part of the IBC.

Section 102.6 of the IBC further references the IEBC, IPMC, and IFC when dealing with existing buildings. Of the additional codes referenced above, it is perhaps most important that users of the IEBC also have a keen understanding of the requirements of the IPMC and IFC. Not only does the IBC reference these codes for existing buildings, but they are referenced in many instances throughout the IEBC.

As an example of what might be included in these additional codes, consider the mechanical equipment located within an existing building. Section 102 of the IMC provides specific requirements for existing mechanical installations; for systems undergoing maintenance; for additions, alterations, or repairs to existing systems; for changes in occupancy; for historic buildings; and for moved buildings. The requirements vary from allowing the existing systems to be utilized as is to requiring new parts to be in accordance with the current code while not requiring the entire system to comply or requiring the installation of new equipment meeting the current IMC.

The design professional must take the time to see how the repairs, alterations, additions, change in occupancy, or moved building are affected by these additional codes and should not rely on the provisions of the IEBC alone. Consider a building that is to be moved. Chapter 14 of the IEBC discusses some of the structural considerations that will need to be made as part of relocating the building, but Section 102.7 of the IMC requires that the mechanical systems within the building comply with the code requirements for new installations.

Building owners and their design professionals should also be aware of those provisions in the IFC that retroactively apply to certain existing buildings and conditions. Chapter 11 of the IFC has provisions for existing buildings and retroactive requirements.

3.2.3 Existing materials. Section 302.4 of the IEBC allows existing materials to remain within the building with two caveats. First, the material in question should have been allowed by the code that was in place at the time of original construction. Second, the code official can require an "unsafe" material to be removed.

Existing buildings often contain hazardous materials such as asbestos, lead, mercury, polychlorinated biphenyls (PCBs), chlorofluorocarbons, and radioactive materials. The Environmental Protection Agency (EPA) provides general oversight in relation to hazardous building materials, while some state and local agencies may have additional requirements (Baldwin, 2016). The following lists some general EPA requirements in relation to some of these hazardous materials:

- *Asbestos:* The EPA requires that an asbestos survey be conducted prior to any demolition. If the survey finds that asbestos-containing materials (ACMs) exist, an abatement plan must be put into place and the EPA must be notified at least 10 days prior to abatement activities.
- *Lead:* The EPA has issued a lead-safe practices rule to help reduce the risk of lead poisoning. The rule is specifically for contractors that perform renovation or repair work that could disturb lead-based paint in homes, childcare facilities, and schools that were built before 1978. The Occupational Safety and Health Administration (OSHA) requires that respiratory protection be used when any lead-based materials might be disturbed.
- *Mercury:* Mercury can be found in lighting, batteries, or mechanical equipment within a building. The EPA regulates mercury as a hazardous waste and requires that any mercury-containing materials be removed prior to demolition or renovation activities.

Prior to commencing any repairs or alterations within an existing building, the owner or contractor should make every effort to identify any hazardous materials within the building. While Section 302.4 of the IEBC states that the code official can require the removal of these materials, this is often regulated by the EPA or a local health department. The owner must determine what the federal, state, and local requirements are and should remove and dispose of suspect materials accordingly.

3.2.4 New and replacement materials. All new construction materials, whether replaced or new elements, are to conform to the requirements of the IBC. The code official should ensure that all new structural elements, as well as their attachments, are designed and detailed to meet the provisions for new construction. Section 302.5 of the IEBC does provide for the following two exceptions to this requirement:

1. "Like" materials are allowed for work involving repairs or alterations so long as an unsafe condition is not created. As an example, consider an unreinforced masonry (URM) building located in a high-seismic region. As part of a minor alteration being

Figure 3.3 Brick re-pointing.

performed, the owner wants to repair the mortar in the exterior brickwork. This is known as "re-pointing" and involves the removal of the old mortar and then the proper installation of new mortar as shown in Fig. 3.3. While the IBC does not allow new URM buildings to be constructed in high-seismic regions, this exception to the code would allow the existing brickwork to be re-pointed. In truth, re-pointing the brick provides additional shear capacity and is often an important aspect to performing a seismic upgrade of a URM building.

2. Where alternate materials, design, or methods are used in accordance with Section 104.11 of the IEBC. An example of an alternate material that could be used is shown in Fig. 3.4. This image depicts rammed earth walls that are quite common in different regions of the United States such as Arizona, New Mexico, and Texas. Rammed earth walls can serve as both bearing and shear walls and require that the specific type and gradation of soils be specified, along with the lift thicknesses and the compaction requirements, as well as any specific requirements for mild reinforcing steel.

3.2.5 Occupancy and use. When using the IEBC, it will often be important to understand how the building or space is being used. Both the IEBC and IBC have specific code provisions that are triggered based upon the use of the space. Section 302.6 of the IEBC simply refers to Chapter 3 of the IBC when determining the use of the space. How the space will be utilized is described by the occupancy classification that is assigned. Table 3.1 lists the occupancy classifications that are described in Chapter 3 of the IBC.

More time will not be spent here describing the differences between these occupancy classifications. To learn more, the reader should spend some time reviewing Chapter 3 of the IBC in its entirety.

Figure 3.4 Rammed earth wall. (*Inhabitat, 2016.*)

Table 3.1 Occupancy Classifications

Types of Use	Occupancy Group	Occupancy Subgroups
Assembly	Group A	A-1, A-2, A-3, A-4, A-5
Business	Group B	None
Educational	Group E	None
Factory and industrial	Group F	F-1, F-2
High-hazard	Group H	H-1, H-1, H-3, H-4, H-5
Institutional	Group I	I-1, I-2, I-3, I-4
Mercantile	Group M	None
Residential	Group R	R-1, R-2, R-3, R-4
Storage	Group S	S-1, S-2
Utility and miscellaneous	Group U	None

3.3 *Structural Requirements*

The IEBC includes many structural triggers that require items to be evaluated to the current code or a variation of the current code. Many times, design professionals or code officials will refer to these as mandatory upgrade triggers, but this is not necessarily correct. There is an important difference between requiring an evaluation versus mandating an upgrade.

For an example, consider that a new 500-lb rooftop unit will be placed on an existing roof. Section 806.2 requires that the existing roof framing that will support the new unit show compliance with the IBC. It could be that the analysis performed shows that the existing construction is adequate and would not require that existing structural elements be replaced or altered to comply.

With that being said, if the proposed work will be significant and involves an older building, one can often assume that it will require an upgrade in order to comply with the current code. While that is often the case, the code official should make it clear that the code is simply requiring an evaluation, or better yet, an analysis that shows compliance with the code. If the analysis identifies deficiencies, an upgrade will be required.

Throughout the IEBC there are numerous structural triggers that will require an evaluation. This makes Section 303 of the IEBC very important as it provides the minimum requirements for structural evaluations. Both the SER and the code official should have a clear understanding of what is required by this section. Section 303 of the IEBC identifies different requirements for evaluation depending on whether the trigger is related to a change in live loads, a change in snow loads, or a specific seismic requirement. The following sections will address the evaluation requirements for each separately.

3.3.1 Live loads. If a live load trigger has been met and the structural components supporting the gravity loads must be evaluated, Section 303.1 of the IEBC allows for three separate paths of compliance. Those are as follows:

- > *Live Load:* Anytime the live load has been increased, the existing framing must be checked for compliance with the current code. As an example, consider an existing building where a portion of one floor is being changed from an office space to a restaurant. Table 1607.1 of the IBC lists a design live load of 50 lb/ft^2 for office space but lists a value of 100 lb/ft^2 for restaurants. In this example, the existing floor framing supporting the restaurant area, and other elements providing support to such framing, will need to be evaluated to ensure compliance with the current code.
- ≤ *Live Load:* If the alterations being made do not cause an increase in the design live loads, or they lessen the design live loads, the existing floor framing and supporting elements can be analyzed per the code at the time of original construction. This would be applicable to most projects that do not undergo a change in occupancy.
- *Noncomforming:* Section 303.1 includes a provision of nonconformity. It allows the code official to approve a lesser live load than what is listed in Table 1607.1 of the IBC. While many code officials may not allow this, if it is allowed then a placard must be provided clearly noting the approved live load. Figure 3.5 provides an example of what this placard might look like.

Figure 3.5 Floor live load placard.

This is not to be confused with the live load posting requirements in the IBC or required by OSHA. Section 106.1 of the IBC requires that floor live loads in commercial or industrial buildings be posted when they exceed 50 lb/ft^2. OSHA, under Section 1926.250(a)(2) of the *Code of Federal Regulations*, requires that the live loads for storage areas be posted unless it is located on a slab-on-grade. The IBC and OSHA posting requirements are significantly different as they are required even when the live load conforms to the code requirements. This exception in the IEBC is for nonconforming applications only.

As an example, assume that a single-family residence is converted into an office. The floor live load for the residence was 40 lb/ft^2 while Table 1607.1 of the IBC requires a floor live load of 50 lb/ft^2 for office space. The owner of the space could provide information to the code official showing that it will be used as an accounting office and will likely not have more than three persons at a time. If that is the case, they could argue that the 40 lb/ft^2 live load is appropriate. If this is allowed by the code official, they will need to post the 40 lb/ft^2 live load in accordance with Section 303.1 of the IEBC.

3.3.2 Snow loads. Section 303.2 of the IEBC requires that existing roof framing members be checked for new snow drift loads that could occur because of the additions or alterations that have been made. The following list highlights three instances when this could occur:

- *Projections:* Many additions or alterations may create a projection that would cause snow drifts to occur. The most common time for this to happen is when an addition is made that is taller than the original structure as shown in Fig. 3.6. The taller addition causes new snow drift loads at the original structure and would require the existing roof framing to be checked for the new loads. Other examples could be adding large rooftop units, skylights, solar panel arrays, or even re-framing a portion of a roof to create a pop-up.
- *Adjacent Building:* As shown in Fig. 3.6, if an addition is constructed near an adjacent building and creates a shade effect on the adjacent building, new snow drift loads could occur at the adjacent roof. This could also occur if a relocated building is placed near an adjacent building. As shown in Fig. 3.6, this only needs to be considered when the adjacent structure is within 20 ft of the addition or the relocated building and less than six times the vertical separation distance.

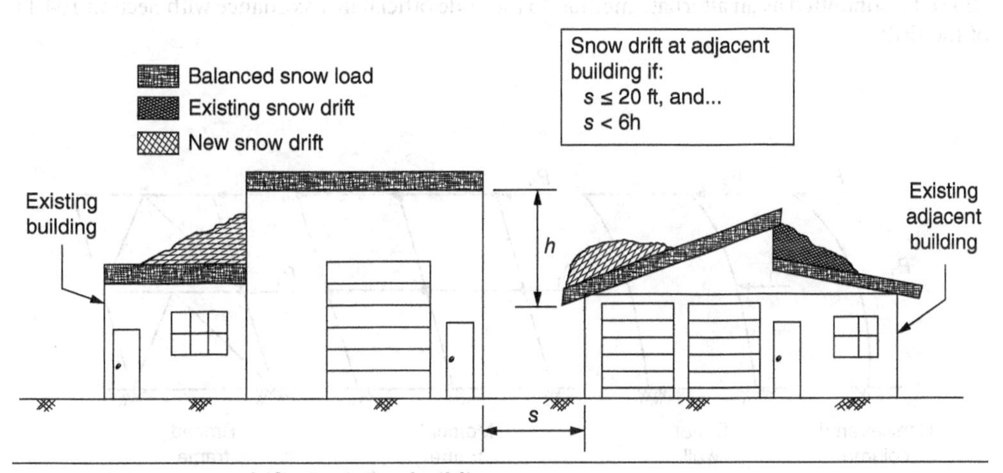

Figure 3.6 New snow drift at existing buildings.

- *Relocated Building:* When a building is moved or relocated to an area with higher snow loads, additional snow drift loads could occur. As noted in Chapter 7 of ASCE 7-16, snow drift loads should be considered at lower roofs, adjacent structures, and roof projections and parapets. If any of these conditions occur in the moved structure, a check of the snow drift loads should be provided.

These new snow drift loads could require significant alterations to the existing roof framing. With that said, snow drifts can affect other aspects of the building design as well, which should be considered by both the designer and the code official. This includes items such as blocking doors, blocking air intake or exhaust openings, blocking windows and causing leakage, burying rooftop equipment, and more. In addition to requiring a structural evaluation of the existing framing members, these items should be considered carefully in light of the new structural loads that are created.

3.3.3 Seismic evaluations. Throughout the IEBC there are many instances where seismic evaluations are required. Section 303.3 of the IEBC outlines the methods in which these evaluations are to be performed and is broken up into two parts. It is imperative that the code officials have a clear understanding of what is required in relation to each of these two parts.

The first part is discussed in Section 303.3.1 and covers when the evaluation must conform to 100 percent of the seismic forces prescribed by the IBC. This is termed "full" compliance. The second part addresses when the evaluations can be performed to a "reduced" seismic performance level. When the seismic evaluation triggers are discussed in the IEBC, it will specifically note whether "full" or "reduced" compliance is required.

For seismic evaluations meeting the full compliance requirement, there are two methods that are allowed by the IEBC. They are described below.

IBC compliance. This requires that existing buildings be evaluated for 100 percent of the seismic forces that are calculated in accordance with Chapter 16 of the IBC. As many existing buildings are quite old and the level of seismic detailing is often not known, Section 303.3.1 requires that the designer assume that the existing seismic force–resisting system (SFRS) be considered "ordinary" in the evaluation.

To understand this requirement, it is first important to understand what the SFRS is. This term typically refers to the vertical seismic-resisting elements only. Figure 3.7 shows the four most common types of vertical SFRSs. Each of these can be built using common structural materials such as wood, concrete, masonry, or steel. Other types of materials or SFRSs may be allowed if submitted as an alternate method to the code official in accordance with Section 104.11 of the IEBC.

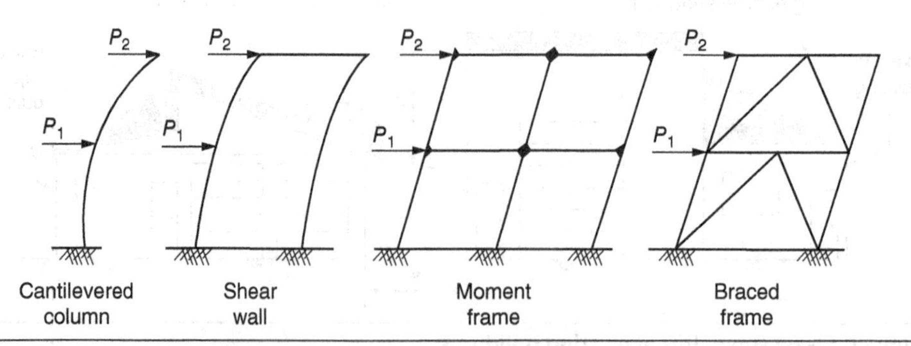

Figure 3.7 Common vertical seismic force–resisting systems.

Now to the limitation provided in Section 303.3.1, that the SFRS must be deemed "ordinary." Not only can a designer choose different materials for each of the SFRSs shown in Fig. 3.7, they can also detail each of these systems in a different way to make them more resistant to lateral forces induced by either seismic or wind events. This detailing is described as either ordinary, detailed, intermediate, or special. A SFRS that is detailed as "ordinary" is the least resistant to lateral forces, while a SFRS detailed as "special" is the most resistant. The IEBC limitation requires that the SER assume that the least resistant system exists unless enough information can be provided to show that a detailed, intermediate, or special system can be considered in the evaluation.

To better understand the difference between SFRSs, consider the design of a new building. Before commencing the design, the SER refers to Table 12.2-1 of ASCE 7-16 for the selection of an appropriate SFRS. This table forms the basis of the seismic design for new construction that is performed by the SER. The first column describes the SFRS being selected, while each subsequent item in the respective row provides the design limitations for that system.

As an example, assume that the SER is designing a new two-story office building in Seismic Design Category D. The owner wants to construct the building using reinforced masonry bearing walls. By selecting system A.9 from Table 12.2-1, ordinary reinforced masonry shear walls, the SER sees that they are not allowed in Seismic Design Category D. To use masonry bearing walls for this high-seismic region, they will need to design them as A.7, special reinforced masonry shear walls, per the same table. Table 12.2-1 provides limitations for when some SFRSs can be used in high-seismic regions, but it also prescribes maximum height limitations in many cases. If it states NL, that means there is no height limitation.

The middle columns within Table 12.2-1 provide the design values that form the basis of the seismic analysis. While each of these factors is important, the response modification coefficient has perhaps the greatest effect on the design. To help understand how the response modification factor works, consider the difference between wood-sheathed shear walls and an ordinary steel moment frame. Table 12.2-1 provides a response modification coefficient of 6.5 for wood-sheathed shear walls (A.15), while a response modification coefficient of 3.5 is listed for ordinary steel moment frames. The higher the coefficient, the greater the resistance of the SFRS to seismic forces. When calculating the seismic forces (e.g., base shear) the ordinary steel moment frames would only have 54 percent of the capacity that the wood-sheathed shear walls would have.

Similarly, comparing the response modification coefficients of an "ordinary" system as compared to a "special" system shows a huge difference in seismic resistance capacity. Consider an ordinary reinforced masonry shear wall (A.9) with a coefficient of 2.0 as compared to a special reinforced masonry shear wall (A.7) with a coefficient of 5.0. The ordinary reinforced masonry shear wall only has 40 percent of the capacity to resist seismic forces as the special reinforced masonry shear walls.

The other two design factors provided in Table 12.2-1 of ASCE 7-16 are the deflection amplification factor (C_d) and the overstrength factor (Ω_0). The deflection amplification factor is used to adjust lateral displacements for the structure. The more ductile a system is, the greater the value of the response modification and deflection amplification factors. The overstrength factor amplifies the seismic forces when certain structural irregularities exist in a building. Forces tend to concentrate in areas of these irregularities so the overstrength factor allows the designer to estimate what these increased forces might be.

In summary, when the IEBC trigger notes that "full" seismic compliance is required, the SER must provide an analysis that conforms to Chapter 16 of the IBC. The seismic design must be performed in accordance with Chapter 12 of ASCE 7-16 as referenced in Section 1613.1 of the IBC. When selecting the SFRS from Table 12.2-1 of ASCE 7-16, the SER must assume an "ordinary" system unless enough information is provided to allow the use of a detailed, intermediate, or special system.

ASCE 41 compliance. In lieu of showing full compliance with the IBC, Section 303.3.1 allows the SER to use ASCE 41-17 for the seismic evaluation. ASCE 41 provides a methodology for providing a seismic evaluation of existing buildings. The standard allows the owner and design professional to select a "performance objective" for the building, and the seismic evaluation process then identifies any deficiencies that may exist to meet that performance objective. ASCE 41 also prescribes four different analysis methods that can be used to perform the retrofit design that would be required to address the deficiencies noted in the evaluation.

While ASCE 41 allows for several different iterations of performance objectives, Section 303.3.1 of the IEBC establishes what those performance objectives should be when "full" compliance with the IBC is required. Table 303.3.1 of the IEBC, as shown in Fig. 3.8, displays the minimum performance objectives that must be met when using ASCE 41 in lieu of the IBC for full compliance.

The ASCE 41 methodology is selected quite often when dealing with existing buildings in high-seismic regions. As such, it is imperative that the code officials have a good understanding of how this standard is laid out and what the information provided in Table 303.3.1 of the IEBC represents. Section 3.4 of this chapter provides more specific information on the ASCE 41 standard.

The above-noted evaluation requirements are needed when "full" IBC compliance is specified. When "reduced" compliance is allowed, there are three separate paths that the design professional can take as described below.

- *Reduced IBC Compliance:* This requires that existing buildings be evaluated considering only 75 percent of the IBC-prescribed seismic forces. Similar to full compliance, the IBC-required forces are calculated in accordance with Chapter 16 of the IBC but are then reduced by 25 percent. This is often the easiest, and most commonly used, path of compliance when conforming to the reduced IBC force level.
    - The field of seismic-strengthening existing buildings has been led by the efforts of structural engineers on the West Coast. Previous to the adoption of the IBC, structural design on the West Coast was performed in accordance with the UBC. Many design professionals and jurisdictions alike felt that analyzing existing buildings to meet the full requirements of the UBC was not necessary. The main reason for this was that existing buildings have a shorter useful life than new buildings. For this reason, the use of 75 percent of the UBC seismic forces became a common practice and has carried over to the IEBC provisions (Dal Pino, 2016).
- *Reduced ASCE 41 Compliance:* Rather than follow the path of the IBC, Section 303.3.2 of the IEBC also allows the use of ASCE 41-17 when performing a seismic evaluation to the reduced IBC force level. As discussed previously, ASCE 41 allows the owner to select from several different performance objectives, but Table 303.3.2 of the IEBC (see Fig. 3.9) prescribes the minimum level of performance required for compliance with the reduced IBC seismic requirements. Much more information in relation to ASCE 41 is provided in Sec. 3.4.

PERFORMANCE OBJECTIVES FOR USE IN ASCE 41 FOR COMPLIANCE WITH FULL SEISMIC FORCES		
RISK CATEGORY (Based on IBC Table 1604.5)	STRUCTURAL PERFORMANCE LEVEL FOR USE WITH BSE-1N EARTHQUAKE HAZARD LEVEL	STRUCTURAL PERFORMANCE LEVEL FOR USE WITH BSE-2N EARTHQUAKE HAZARD LEVEL
I	Life Safety (S-3)	Collapse Prevention (S-5)
II	Life Safety (S-3)	Collapse Prevention (S-5)
III	Damage Control (S-2)	Limited Safety (S-4)
VI	Immediate Occupancy (S-1)	Life Safety (S-3)

Figure 3.8 Table 303.3.1 of the IEBC. ***(Excerpted from the 2018 International Existing Building Code; copyright 2017. Washington, D.C.: International Code Council.)***

PERFORMANCE OBJECTIVES FOR USE IN ASCE 41 FOR COMPLIANCE WITH REDUCED SEISMIC FORCES		
RISK CATEGORY (Based on IBC Table 1604.5)	STRUCTURAL PERFORMANCE LEVEL FOR USE WITH BSE-1E EARTHQUAKE HAZARD LEVEL	STRUCTURAL PERFORMANCE LEVEL FOR USE WITH BSE-2E EARTHQUAKE HAZARD LEVEL
I	Life Safety (S-3). See Note a	Collapse Prevention (S-5)
II	Life Safety (S-3). See Note a	Collapse Prevention (S-5)
III	Damage Control (S-2). See Note a	Limited Safety (S-4). See Note b
IV	Immediate Occupancy (S-1)	Life Safety (S-3). See Note c
a. For Risk Categories I, II and III, the Tier 1 and Tier 2 procedures need not be considered for the BSE-1E earthquake hazard level. b. For Risk Category III, the Tier 1 screening checklists shall be based on the Collapse Prevention, except that checklist statements using the Quick Check provisions shall be based on *MS*-factors that are the average of the values for Collapse Prevention and Life Safety. c. For Risk Category IV, the Tier 1 screening checklists shall be based on Collapse Prevention, except that checklist statements using the Quick Check provisions shall be based on *MS*-factors for Life Safety.		

Figure 3.9 Table 303.3.2 of the IEBC. (*Excerpted from the 2018 International Existing Building Code; copyright 2017. Washington, D.C.: International Code Council.*)

- *IEBC Appendices:* Section 303.3.2 of the IEBC also allows certain procedures that are provided as an appendix to the IEBC to be used. It is important to understand that these procedures are only allowed when the trigger noted in the IEBC allows compliance with the "reduced" IBC seismic force level. These provisions cannot be used when compliance to the "full" IBC is required.

  The specific procedures referenced in this section are included in Appendix A of the IEBC and are subdivided into Chapters A1 through A4. The following list provides the title of each chapter. The title describes what types of buildings are addressed within the specific provisions included therein.

  - *Chapter A1:* Seismic Strengthening Provisions for Unreinforced Masonry Bearing Wall Buildings
  - *Chapter A2:* Earthquake Hazard Reduction in Existing Reinforced Concrete and Reinforced Masonry Wall Buildings with Flexible Diaphragms
  - *Chapter A3:* Prescriptive Provisions for Seismic Strengthening of Cripple Walls and Sill Plate Anchorage of Light, Wood-Frame Residential Buildings
  - *Chapter A4:* Earthquake Risk Reduction in Wood-Frame Residential Buildings with Soft, Weak or Open Front Walls

This section of this handbook has described what is required when a structural evaluation is triggered for live loads, snow loads, and seismic resistance. Throughout the IEBC there will also be triggers that will require a structural evaluation due to increased dead loads, flood loads, or wind loads. While Chapter 3 of the IEBC does not specifically address the evaluation requirements in these instances, specific information on how those should be provided will be addressed in this book, as those triggers are discussed in future chapters.

3.4 *ASCE 41-17*

In 2010 the Structural Engineers Association of California (SEAOC) performed a survey that was sent out to more than 4000 members of their organization. The purpose of the survey was to get their feedback on the use of ASCE 41. Several interesting items came to light as a result of the survey. Perhaps one of the most important findings was that 82 percent of the respondents stated that there is a need for better-informed code officials in relation to the requirements of ASCE 41 (Maison et al., 2010). While this survey was performed in 2010, the results of the survey would likely be the same if taken today.

The following is a simple summary of how ASCE 41 is laid out. The purpose of this summary is to provide the code official with a general understanding of how the standard is formatted and what they should be looking for when a seismic evaluation is required by the IEBC and the ASCE 41 compliance path is selected. This summary is not intended to assist the SER in learning how to use ASCE 41 to perform seismic evaluations and rehabilitation designs. Many design professionals may not be familiar with ASCE 41 and could easily be intimidated by its 400 pages that include different terminologies, concepts, and procedures from what is included in the IBC (Maison et al., 2010). With that said, it is expected that the SER should have a clear understanding of the ASCE 41 requirements before proceeding with the evaluation and rehabilitation design.

ASCE 41 combines the requirements for the seismic evaluation and seismic rehabilitation design into one standard. Previously the evaluation requirements were included in ASCE 31, and the rehabilitation design was provided in ASCE 41. Both of those standards were developed from "prestandard" guidelines developed by FEMA, such as FEMA 310 and FEMA 356. Each of these "prestandards" was based upon other FEMA and ANSI documents. This is important to understand, as the majority of the provisions included in ASCE 41 have been used for decades and on thousands of existing buildings throughout the United States.

ASCE 41 prescribes three tiers that can be used when performing a seismic evaluation. The tier that is selected depends upon the performance objectives that are selected by the owner and SER as well as the building type that is being evaluated. Once the seismic evaluation has been performed and the deficiencies have been identified, ASCE 41 prescribes two different tiers for the seismic rehabilitation design. To best grasp the requirements of ASCE 41, it is important that the code official understands how the performance objectives are selected, the difference between the three evaluation tiers, and the difference between the two rehabilitation design tiers.

3.4.1 Performance objective. ASCE 41 allows the building owner and SER to select the level of performance they desire the existing building to meet. There are numerous combinations of objectives that they can choose from to develop the performance criteria that the building will be evaluated to. The performance objective takes into account the safety of the occupants during and after an earthquake, the length of time the building is taken out of service, and the economic effects on the larger community.

It is important to understand that the ASCE 41 standard defines the different performance objectives but does not mandate which ones are to be used. With that said, Section 303.3 of the IEBC takes much of the decision-making process away from the owner and SER and outlines the minimum performance objectives that are to be used in the seismic evaluations of existing buildings.

When selecting the performance objective using ASCE 41, there are three important criteria that need to be considered. Those are (1) the risk category of the building in question, (2) the target building performance level, and (3) the seismic hazard caused by ground shaking. Each of these items is noted in Figs. 3.8 and 3.9 (Tables 303.3.1 and 303.3.2 of the IEBC). The following is a brief description of what each of these items refers to.

Risk category. The risk category for the building is outlined in Section 1604.5 of the IBC and is best understood when reviewing Table 1604.5 of the IBC. In general, the risk category for a building increases as the number of persons affected by the building increases. While most buildings are defined as Risk Category II, a large assembly occupancy that has more than 300 persons is required to meet the Risk Category III criteria. Hospitals that must be in operation after significant natural disasters are required to meet the Risk Category IV criteria. As the risk category increases, so do the design wind, seismic, and snow loads on the building.

It is important that the code officials have a clear understanding of what risk category a building should be assigned. As the risk category increases, so do the performance objectives for the seismic evaluation. Further discussion will not be provided at this time in relation to the risk category. The code official should take the time necessary to review Table 1604.5 of the IBC to gain a better understanding of what building types fall into which risk category.

Target building performance level. Section 2.3.1 of ASCE 41 outlines the five performance levels that are typically considered. Those include immediate occupancy (S-1), damage control (S-2), life safety (S-3), limited safety (S-4), and collapse prevention (S-5). Design procedures and acceptance criteria for each of these performance levels are provided throughout the standard. The performance levels that are most important to understand are the immediate occupancy, life safety, and collapse prevention levels. The following is a brief description of what each means:

- *Immediate Occupancy (S-1):* If a structure complies with the immediate occupancy condition, it means that after the design-level earthquake the structure should still be safe to occupy. If damage does occur, it should be limited and should not affect the main SFRS or gravity-resisting members. The risk of any life-threatening injuries is considered to be very low.
- *Life Safety (S-3):* A structure design to the life safety performance level is expected to experience severe damage under the design-level earthquake, but a margin of safety should still be provided against total or partial collapse of the structure. Injuries might occur, but the overall risk of life-threatening injuries is considered to be low. In many cases it may be possible to repair the structure.
- *Collapse Prevention (S-5):* When evaluated to the collapse prevention level very little safety margin is provided. After the design-level earthquake a structure evaluated to this level is on the verge of total or partial structural collapse. It suggests that the structure will still be able to support gravity loads, but repair of the building will likely not be possible. There also might be risk of severe injuries from falling debris, but all occupants should be able to exit the building.

Table C2-4 of ASCE 41 provides examples of the level of damage that would be seen for each of the performance levels noted above based upon the building type. As an example, consider a steel moment–frame building. If the building were evaluated to the collapse prevention under the design-level earthquake, the frames would likely see extensive distortion, and fractures in several moment connections would likely be found, while shear connections should remain intact. If the same building were evaluated to the life safety performance level, there will likely be some hinges that form, some severe joint distortion, and some local buckling, but the shear connection will remain intact. When considering the immediate occupancy performance level, minor yielding may occur in a few locations but no fractures should occur.

Level S-2, or the "damage control" performance level, is taken halfway between the life safety and immediate occupancy performance levels. Similarly, level S-4 is taken as halfway between the collapse prevention and life safety performance levels. When reviewing Tables 303.3.1 and 303.3.2 of the IEBC, one can see that the S-2 and S-4 performance levels are only considered when dealing with Risk Category III structures. Figure 3.10 provides a visual of the difference between these five performance levels.

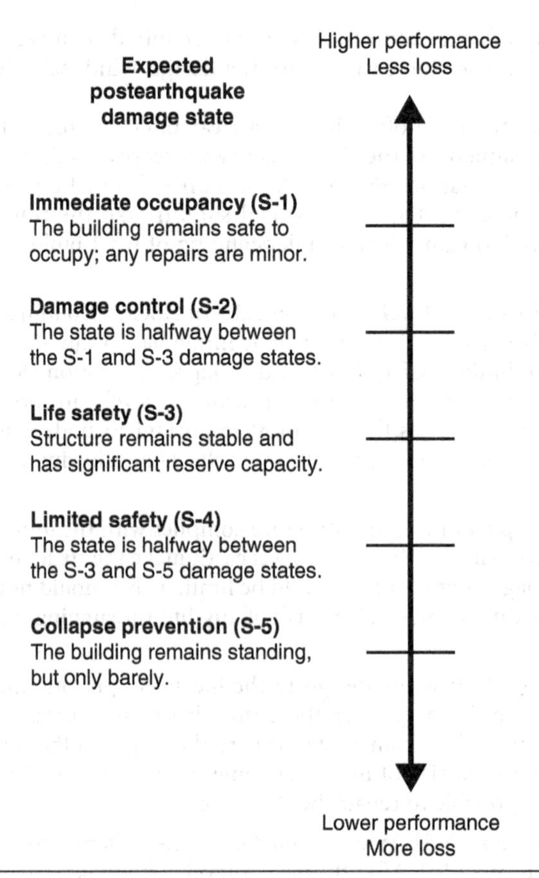

Figure 3.10 ASCE 41 target building performance levels.

Seismic hazard. The selected seismic hazard is a function of the proximity of the building to active faults, the magnitude and return frequency of seismic events, and the soil conditions at the site. ASCE 41 defines four separate seismic hazard levels that can be selected as part of the evaluation. Those are as follows:

- *BSE-2N:* This is termed the maximum considered earthquake (MCE_R) and represents a seismic event that has a 2-percent probability of exceedance in 50 years. That is equivalent to the maximum earthquake that would be experienced at a site having a mean return period of 2475 years.
- *BSE-1N:* This is termed the design-level earthquake and considers ground motions that are two-thirds of the values considered in the BSE-2N response acceleration. The design-level earthquake has a 10-percent probability of exceedance in 50 years, which is equivalent to a mean return period of 475 years.
- *BSE-2E:* The BSE-2E was developed specifically for existing buildings and roughly 75 percent of the BSE-2N MCE_R event. It has a 5-percent probability of exceedance in 50 years, which is equivalent to a mean return period of 975 years.
- *BSE-1E:* Similar to the BSE-1N, the BSE-1E is considered the design-level event and is equal to two-thirds of the BSE-2E ground motions. It is equivalent to a 20-percent probability of exceedance in 50 years, or otherwise has a mean return period of 225 years.

Note that the first two seismic hazards end with the letter "N" while the last two end with the letter "E." The "N" signifies new construction, or compliance with the "full" IBC seismic hazards. The "E" signifies existing construction, or compliance with the "reduced" IBC seismic hazards.

Now that the risk category, target building performance level, and seismic hazards are defined, they need to be put together to come up with the building's performance objective. To understand this, please review Fig. 3.11, which is a reproduction of Table 2-2 of ASCE 41. This is considered the Basic Performance Objective Equivalent to New Building Standards, or BPON. When comparing this table to Table 303.3.2 of the IEBC (Fig. 3.9), the reader will notice that they include the same requirements. As such, when evaluating a building for compliance with the "full" IBC seismic requirements, it would need to comply with the provisions of this table. Figure 3.11 shows that the building performance objective varies depending upon the risk category of the structure. It also shows that the evaluation needs to consider the building performance during the BSE-1N and BSE-2N seismic hazard events. This evaluation must follow the Tier 3 procedures described in Chapter 6 of ASCE 41. To understand how to use this table, consider that an existing fire station is being evaluated to the BPON, or "full" seismic requirements. As a Risk Category IV structure, it will need to be checked for immediate occupancy under the design-level event (BSE-1N) and for life safety during the MCE_R (BSE-2N) event.

Basic Performance Objective Equivalent to New Building Standards (BPON)		
	Seismic Hazard Level	
Risk Category	BSE-1N	BSE-2N
I & II	Life Safety Structural Performance Position Retention Nonstructural Performance (3-B)	Collapse Prevention Structural Performance Nonstructural Performance Not Considered (5-D)
III	Damage Control Structural Performance Position Retention Nonstructural Performance (2-B)	Limited Safety Structural Performance Nonstructural Performance Not Considered (4-D)
IV	Immediate Occupancy Structural Performance Operational Nonstructural Performance (1-A)	Life Safety Structural Performance Nonstructural Performance Not Considered (3-D)

Figure 3.11 **Table 2-2 of ASCE 41. (*With permission from ASCE.*)**

Figure 3.12 shows Table 2-1 of ASCE 41. This table displays what is called the Basic Performance Objective for Existing Buildings, or BPOE. Notice that this table is similar to Table 2-2 in that the requirements are based upon risk category, but there are a couple of differences that need to be highlighted. First, it allows the user to perform the evaluation using any of the three evaluation tiers. Second, the Tier 1 and Tier 2 requirements only consider a single seismic hazard, while the Tier 3 procedure requires that two seismic hazards be considered, as with Table 2-2.

If the user were to cross out the Tier 1 and Tier 2 portions of the table and compare it to Table 303.3.2 of the IEBC (Fig. 3.9), they would find that the requirements are the same. As such, Table 303.3.2 of the IEBC is based on the Tier 3 requirements of the BPOE. Similarly, the BPOE can be compared to the "reduced" seismic provisions of the IBC.

Basic Performance Objective for Existing Buildings (BPOE)				
	Tier 1[a]	Tier 2[a]	Tier 3	
Risk Category	BSE-1E	BSE-1E	BSE-1E	BSE-2E
I& II	Life Safety Structural Performance Life Safety Nonstructural Performance (3-C)	Life Safety Structural Performance Life Safety Nonstructural Performance (3-C)	Life Safety Structural Performance Life Safety Nonstructural Performance (3-C)	Collapse Prevention Structural Performance Nonstructural Performance Not Considered (5-D)
III	See footnote *b* for Structural Performance Position Retention Nonstructural Performance (2-B)	Damage Control Strucrural Performance Position Retention Nonstructural Performance (2-B)	Damage Control Structural Performance Position Retention Nonstructural Performance (2-B)	Limited Safety Structural Performance Nonstructural Performance Not Considered (4-D)
IV	Immediate Occupancy Structural Performance Position Retention Nonstructural Performance (1-B)	Immediate Occupancy Structural Performance Position Retention Nonstructural Performance (1-B)	Immediate Occupancy Structural Performance Position Retention Nonstructural Performance (1-B)	Life Safety Structural Performance Nonstructural Performance Not Considered (3-D)

[a]For Tier 1 and 2 assessments, seismic performance for the BSE-2E is not explicitly evaluated.

[b]For Risk Category III, the Tier 1 screening checklists shall be based on the Life Safety Performance Level (S-3), except that checklist statements using the Quick Check procedures of Section 4.5.3 shall be based on MS-factors and other limits that are an average of the values for Life Safety and Immediate Occupancy.

Figure 3.12 Table 2-1 of ASCE 41. (*With permission from ASCE.*)

If comparing the tables during the discussion, the reader has likely noticed another significant difference between the IEBC and ASCE 41 tables. While ASCE 41 prescribes requirements for the evaluation of structural and nonstructural elements of the building, the IEBC performance objectives listed in Section 303 only require the structural elements to be evaluated. The ASCE 41 evaluation procedures make a considerable effort to also address the restraint of nonstructural components as they have historically caused most of the monetary damage to existing buildings in the United States during seismic events. It is likely that future versions of the IEBC will also focus more on this requirement to address the restraint of nonstructural components in regions of high seismicity.

3.4.2 Evaluation tiers. ASCE 41 prescribes three different levels of seismic evaluations, or "tiers." Section 1.4.4 of ASCE 41 states that the tier that is selected depends upon the selected performance objective, the level of seismicity, and the building type. With that said, the code official must be aware that if a "full" or "reduced" seismic evaluation is triggered by the IEBC, only a Tier 3 evaluation is allowed. Tier 1 and Tier 2 are known as "deficiency-based" evaluations and can only be used when the building is undergoing a voluntary seismic retrofit as will be explained later in this book.

Deficiency-based procedures (Tier 1 and Tier 2) require a less detailed evaluation in order to check compliance with a particular performance objective. They represent a good method to identify potential deficiencies that exist in the structure. While Section 303 of the IEBC specifically requires a Tier 3 evaluation when "full" or "reduced" seismic requirements are triggered, the ASCE 41 standard itself does not make this distinction and will allow the SER to select a Tier 1 or Tier 2 procedure anytime certain criteria are met. Because the SER may go directly to ASCE 41 and may not rely on the requirements of the IEBC, it is important that the code officials make the distinction between each of these three tiers and understand the minimum requirements of a Tier 3 evaluation.

Tier 1 and Tier 2 procedures can only be used if the structure in question complies with one of the "common building types" provided in Table 3-1 of the standard, with some limitations as to mixing types. The building types addressed encompass most types of buildings and include wood light-framed structures, steel moment frames, steel braced frames, dual frame systems, steel light frames, steel frames with infill masonry walls, concrete moment frames, precast or tilt-up concrete, reinforced masonry bearing walls, URM bearing walls, and steel plate shear walls.

In addition to limiting the building type, the Tier 1 and Tier 2 evaluations are subject to the boundaries set in Table 3-2 of the standard (see Fig. 3.13). This table prescribes the maximum number of stories a common building type can have when using the Tier 1 or Tier 2 procedures. If the building has more stories than is listed, the Tier 3 procedure is required. In addition, the table notes that the Tier 1 and Tier 2 procedures are not permitted ("NP") in four other instances. The first two are in relation to precast concrete frames and URM buildings in moderate-to high-seismic regions. These types of buildings require the Tier 3 procedure to be used when evaluating to the immediate occupancy (S-1) performance level. The other two instances are for buildings using steel plate shear walls or seismic isolation systems for which Tier 1 and Tier 2 deficiency-based procedures have yet to be developed. The following provides a brief explanation of each tier described in ASCE 41:

- *Tier 1:* The Tier 1 procedure is known as the "screening" phase of the seismic evaluation. While the Tier 1 procedure cannot be used to design the seismic retrofit, it is a good starting point for the identification of potential structural and nonstructural deficiencies that may exist. Chapter 4 of ASCE 41 walks the design professional through the steps of performing a Tier 1 evaluation.

Limitations on the Use of the Tier 1 and Tier 2 Procedures

	Number of Stories[b] Beyond which the Tier 3 Systematic Procedures Are Required							
	Level of Seismicity							
	Very Low		Low		Moderate		High	
Common Building Type[a]	S-3	S-1	S-3	S-1	S-3	S-1	S-3	S-1
Wood Frames								
Light (W1)	NL	NL	NL	4	4	4	4	4
Multi-story, multi-unit residential (W1a)	NL	NL	NL	6	6	6	6	4
Commercial and industrial (W2)	NL	NL	NL	6	6	6	6	4
Steel Moment Frames								
Rigid diaphragm (S1)	NL	NL	NL	12	12	8	8	6
Flexible diaphragm (S1a)	NL	NL	NL	12	12	8	8	6
Steel Braced Frames								
Rigid diaphragm (S2)	NL	NL	NL	8	8	8	8	6
Flexible diaphragm (S2a)	NL	NL	NL	8	8	8	8	6
Steel Light Frames (S3)	NL	1	1	1	1	1	1	1
Dual Systems with Backup Steel Moment Frames (S4)	NL	NL	NL	12	12	8	8	6
Steel Frames with Infill Masonry Shear Walls								
Rigid diaphragm (S5)	NL	NL	NL	12	12	8	8	4
Flexible diaphragm (S5a)	NL	NL	NL	12	12	8	8	4
Steel Plate Shear Wall (S6)	NP[c]	NP[c]	NP[c]	NP[c]	NP[c]	NP[c]	NP[c]	NP[c]
Concrete Moment Frames (C1)	NL	NL	NL	12	12	8	8	6
Concrete Shear Walls								
Rigid diaphragm (C2)	NL	NL	NL	12	12	8	8	6
Flexible diaphragm (C2a)	NL	NL	NL	12	12	8	8	6
Concrete Frame with Infill Masonry Shear Walls								
Rigid diaphragm (C3)	NL	NL	NL	12	12	8	8	4
Flexible diaphragm (C3a)	NL	NL	NL	12	12	8	8	4
Precast or Tilt-Up Concrete Shear Walls								
Flexible diaphragm (PC1)	NL	NL	3	2	2	2	2	2
Rigid diaphragm (PC1a)	NL	NL	3	2	2	2	2	2
Precast Concrete Frames								
Willl shear walls (PC2)	NL	NL	NL	6	6	NP	4	NP
Without shear walls (PC2a)	NL	NL	NL	6	6	NP	4	NP
Reinforced Masonry Bearing Walls								
Flexible diaphragm (RM1)	NL	NL	NL	8	8	8	8	6
Rigid diaphragm (RM2)	NL	NL	NL	8	8	8	8	6
Unreinforced Masonry Bearing Walls								
Flexible diaphragm (URM)	NL	NL	6	4	6	NP	4	NP
Rigid diaphragm (URMa)	NL	NL	6	4	6	NP	4	NP
Seismic Isolation or Passive Dissipation	NP[c]	NP[c]	NP[c]	NP[c]	NP[c]	NP[c]	NP[c]	NP[c]

NOTE: The Tier 3 systematic procedures are required for buildings with more than the number of stories listed herein.
[a]Common building types are defined in Section 3.2.1.
[b]Number of stories shall be considered as the number of stories above lowest adjacent grade.
NL = No Limit (No limit on the number of stories).
NP = Not Permitted (Tier 3 systematic procedures are required).
[c]No deficiency-based procedures exist for these building types. If they do not meet the Benchmark Building requirements, Tier 3 systematic procedures are required.

Figure 3.13 Table 3-2 of ASCE 41. (*With permission from ASCE.*)

- *Tier 2:* The Tier 2 procedure is more involved than the Tier 1 process and includes the retrofit design for smaller projects that do not require a Tier 3 analysis. This procedure is less complicated than the complete analytical evaluation and retrofit requirements outlined in the Tier 3 procedure. The Tier 2 procedure is known as a "deficiency-based evaluation and retrofit," and the steps required are outlined in Chapter 5 of ASCE 41.
- *Tier 3:* Section 3.3.1 of ASCE 41 states that the Tier 3 procedure can be used for any performance level and for any building type. Chapter 6 of ASCE 41 outlines the requirements for the Tier 3 procedure. This procedure is known as the "systematic evaluation and retrofit" method and is much more involved than the Tier 2 path. As required by Section 303.3.1 of the IEBC, when "full" seismic compliance is triggered, the analysis provided must comply with the Tier 3 requirements.

3.4.3 Rehabilitation tiers. Section 1.5.5 of ASCE 41 prescribes either Tier 2 or Tier 3 retrofit procedures depending upon the selected performance objective, level of seismicity, and building type. Tier 2 is considered a deficiency-based procedure, while Tier 3 is termed a systematic retrofit procedure. The procedures for each are defined in Section 3.3 of ASCE 41, with more information in relation to each provided in the ASCE 41 commentary. It is important to note that Section 303.3.1 of the IEBC requires that the Tier 3 procedure can be used when "full" IBC seismic compliance is required.

The ASCE 41 standard allows the SER to select from four separate analysis procedures. These procedures are known as the linear static procedure, linear dynamic procedure, nonlinear static procedure, and nonlinear dynamic procedure (NDP). While the nonlinear procedures are more complicated and require more expertise, they have proven to be significantly less conservative than the linear procedures. The linear procedures are often considered to be excessive, as they have been found to require close to double what is required by evaluations using the IBC in some instances (Dal Pino, 2016). While nonlinear approaches may be preferred, they are quite complicated and may require the code official to seek the assistance of a third-party peer reviewer as discussed in Sec. 3.4.6 below.

The items noted above provide a brief synopsis of how ASCE 41 works when performing a seismic evaluation and subsequent seismic rehabilitation of an existing building. In addition to these items, there are other items noted within ASCE 41 that would be very beneficial to the code official. The following subsections describe items that the code official may need to rely upon when the permit applicant chooses to follow the ASCE 41 path for compliance when an evaluation is triggered in the IEBC.

3.4.4 Evaluation report requirements. Section 1.4.5 of ASCE 41 describes what must be included in the seismic evaluation report and provides much greater detail than is specified in the IEBC. ASCE 41 requires that the evaluation report include the following information at a minimum:

- *Scope and Intent:* This should describe the purpose for the evaluation and the level of investigation that was conducted. If there are any specific requirements of the local jurisdiction or from the IEBC, they should also be listed.
- *Site and Building Data:* This is similar to the code analysis that is often required for new buildings. At a minimum, it is helpful to list the use of the buildings and type of construction. In addition to these items, the following must be listed, as they are key elements in the seismic evaluation performed using ASCE 41:
    - General building description
    - Structural system description
    - Nonstructural system description

- Common building type
- Performance level
- Level of seismicity
- Soil type

- *List of Assumptions:* Any assumptions that were considered must be clearly noted. This includes items such as material properties and any site soil conditions. The code official should pay special attention to the assumptions that are listed. As an example, consider a URM building. One of the most important items to be evaluated is the connection of the roof diaphragm to the exterior bearing walls. This connection is key to preventing collapse and allowing the occupants to escape the building. If the evaluation report states that the roof-to-wall connection could not be inspected but that it was assumed to be adequate, the code official should not accept such an assumption.
- *Findings:* When following the seismic evaluation procedures of ASCE 41, the SER will follow a prescribed format that will help to identify seismic deficiencies in the existing building. The finding portion of the report should clearly list what these deficiencies are and their significance. Before proceeding with a seismic rehabilitation design, it may be beneficial for the owner, SER, and code official to sit down and discuss the findings from the evaluation report. It may be that some items will require resolution while some do not. The code official should be included in this discussion.

The commentary to ASCE 41 also includes several other suggested items that should be included in the evaluation report. Some of those items include the level of inspections and testing that was conducted, the availability of as-built or original design documents, the historical significance of the structure in question, past performance of similar buildings in seismic events, and recommended mitigation schemes. The commentary also suggests that additional items such as references, preliminary calculations, and evaluation checklists be provided as an appendix to the report.

3.4.5 Special inspections and testing. Section 1.5.9 of ASCE 41 states that the construction documents must include requirements for construction quality assurance. Section 1.5.10.1 then highlights the requirement for a quality assurance plan. This terminology is the same as was used in the early versions of the IBC, as the 2000 and 2003 IBC required a quality assurance plan. Since the development of the 2006 IBC, the term has changed to the "Statement of Special Inspections."

Section 1704.3.1 of the IBC lists what needs to be included in the Statement of Special Inspections. This includes noting what items require special inspections and material tests, the extent of those inspections and tests, and the frequency of such inspections and tests. The commentary in ASCE 41 notes that this should also note the contractor quality control procedures, how the SER will be involved in the special inspection program, and any structural observation requirements.

While Chapter 17 of the IBC describes the special inspection and material testing requirements for most common construction materials, many of the materials and construction methods used in relation to existing buildings are not covered. Section 1705.1.1 of the IBC allows the code official to require special inspections and tests for other items not discussed within Chapter 17. This is typically required for alternate materials and systems, unusual design applications, or when specifically required by the manufacturer.

There are many items that may require special inspections in relation to existing buildings that are not specifically noted in Chapter 17 of the IBC. The following provides a sampling of what some of those items might be:

Figure 3.14 FRP composite strengthening. (*Courtesy of Fyfe Tyfo® Fibrwrap®.*)

- *Fiber-Reinforced Polymers (FRPs):* FRP products are often used to strengthen existing concrete or masonry structural elements. Figure 3.14 shows an FRP product manufactured by Fyfe Co., LLC, being applied to a concrete column. Most FRP products should have an evaluation report showing that they comply with the intent of the code. International Code Council® (ICC®) Evaluation Report Number ESR-2103 is specifically for the Tyfo® Fibrwrap® product manufactured by Fyfe Co. This evaluation report lists any limitations of the product, as well as the proper installation procedures. Section 4.3 of this report lists the specific special inspection requirements for the product. In general, FRP products should have third-party special inspections to ensure that the proper materials are being used, the concrete or masonry surfaces have been prepared adequately, the FRP is applied correctly, proper curing is provided, and that any necessary repairs to the FRP have been made.
- *Re-pointing Masonry Joints:* Section 3.2.4 of this chapter discussed the benefits of re-pointing brick mortar joints. In older masonry structures, the mortar can often break down and lose strength. Re-pointing the masonry joint not only improves the physical appearance of the brickwork but also helps to improve the structural integrity of the masonry. If this is deemed to be an integral part of the seismic rehabilitation work, a special inspector should be required to ensure that the old mortar is properly removed, that the surfaces have been cleaned, that the new mortar is of the proper type, and that it has been properly keyed into the joints.

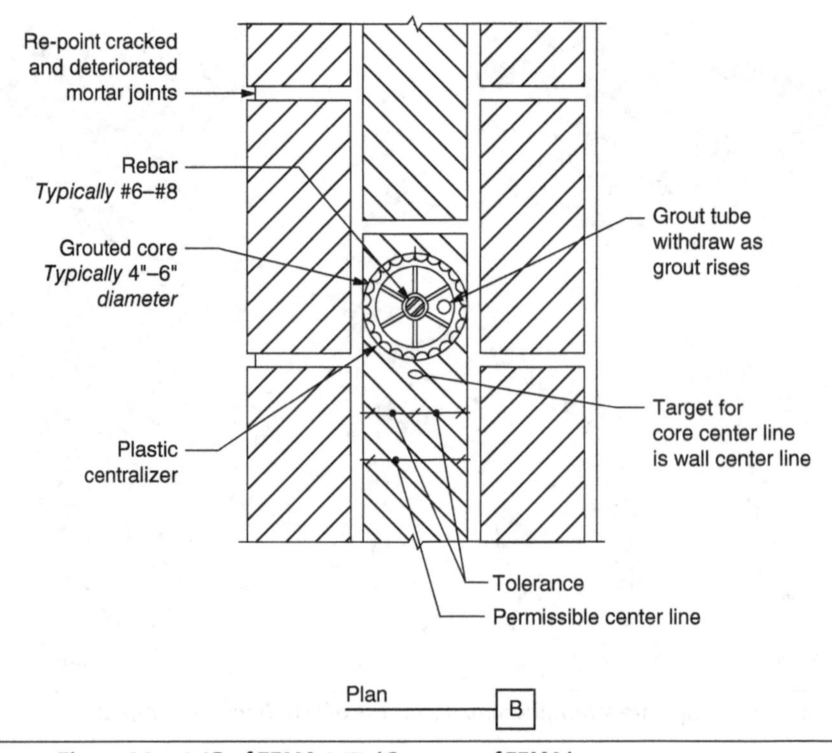

Figure 3.15 Figure 21.4.4-1B of FEMA 547. (*Courtesy of FEMA.*)

- *URM Center Coring:* A common method of strengthening URM bearing walls is to add reinforced cores to the masonry walls. This is done by drilling a core from the roof down the inside of the masonry wall as shown in Fig. 3.15 (FEMA 547, 2006). A steel reinforcing bar is placed in the center of the core, and then it is grouted in place. Third-party special inspections should be provided to ensure that the drilling operations do not cause damage to the surrounding URM; to verify grout strength; to ensure that appropriate reinforcing steel size, grade, and type are being used; and to ensure that centralizers are being used to keep the reinforcement in the center of the core prior to grouting.
- *Postinstalled Anchors:* Many seismic rehabilitation projects will require the addition of postinstalled anchorages. While Table 1705.3 of the IBC notes special inspection requirements for postinstalled anchors, this does not address many of the applications that will occur in existing buildings. Figure 3.16 shows a common postinstalled anchor scenario used to seismically retrofit URM buildings. The anchorage shown in this figure is often specified to help tie the roof or floor framing to the exterior walls, and it is an important connection to limit the potential for structural collapse during a seismic event. Special inspections associated with these anchorages should verify that the hole is properly drilled and cleaned out, that the adhesive materials or postinstalled anchor is being used in the appropriate application, and that the installation complies with the approved construction documents and sometimes includes testing of some of the anchorages. A key consideration when drilling into URM is to use rotary drilling tools rather than using percussion, or impact, tools. If this is specified by the SER, then the special inspector should verify that it was done properly.

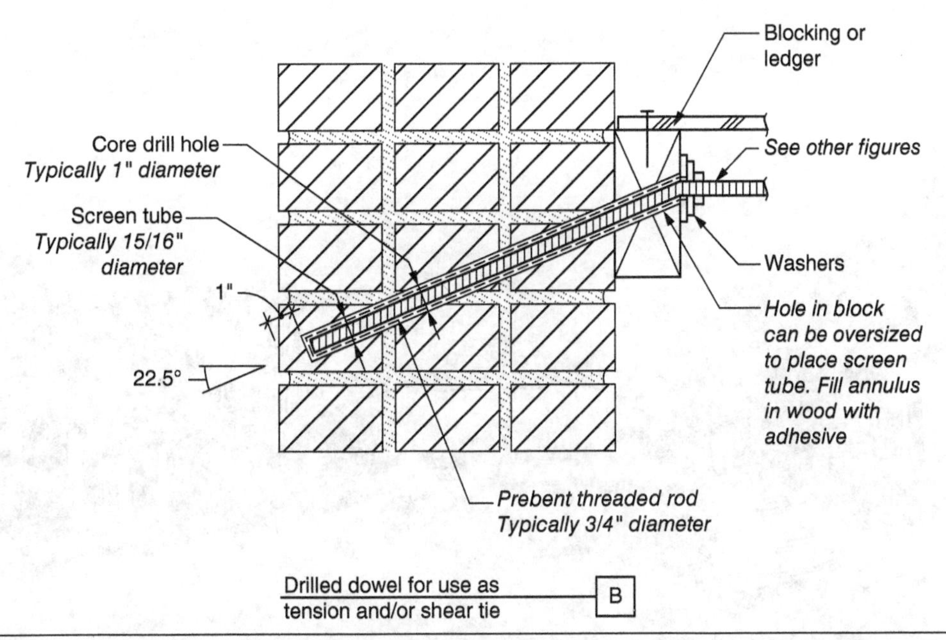

Figure 3.16 Figure 21.4.2-1B of FEMA 547. (*Courtesy of the Federal Emergency Management Agency [FEMA].*)

- *Adjoining Buildings:* In many of the main streets throughout the United States, it is quite common for buildings to adjoin each other and share common walls. The image shown in Fig. 3.17 is taken in Golden, Colorado, but a similar scenario can be found in many places throughout the country. The City of New York has had numerous incidents involving adjoining buildings when work is being to the adjacent building. They have determined that the majority of these incidents is due to one or more of the following items (Eschenasy et al., 2016):
  - *Removal of Support:* This could mean that the existing building is demolished and therefore the support that it provided has been removed from the adjoining building. It could also mean that existing footings become undermined due to the removal of soil or that the groundwater has been lowered significantly and adversely affected the adjoining building's foundations.
  - *Applying New Forces:* The work performed on the building could cause new snow drift loads on the adjoining building, additions or alterations could impose new loads to the common wall, or new foundation loads could pose additional pressures to the adjoining building foundations.
  - *Applying Movement:* Often construction activities involving an existing building can cause excessive vibrations that cause damage to adjacent construction. It could be that new mechanical equipment, such as turbines, is installed within the building, and they in turn cause vibrations that affect adjoining buildings.
  - New York City requires that preconstruction surveys be performed to map existing deficiencies before the work commences, and they also require monitoring during construction to ensure that the work performed does not adversely affect the adjoining building. If work is to occur within an adjoining building, the code official may want to require similar measures be included as part of the Statement of Special Inspections.

Figure 3.17 Main street, Golden, Colorado.

- *Micropile Foundations:* Often, seismic rehabilitation work requires that existing foundations be strengthened. One of the most common methods of doing this is to underpin the existing foundation using micropiles. Micropiles are defined in Section 202 of the IBC as a bored, grouted-in place deep foundation element. They consist of a central steel reinforcing bar and high-strength grout and typically have diameters between 3 and 10 in.
    - While Chapter 17 of the IBC does include special inspection requirements for cast-in-place or driven deep foundation elements, they are not necessarily specific to micropile foundations. Additional requirements that are not included in the IBC could include ensuring that casing is provided if caving soils exist, verify mill certification or test coupon data for reinforcing steel, and verify that pile cap anchorages are provided in accordance with the approved construction documents. Section 1704.19 of the *New York City Building Code* includes additional special inspection requirements for underpinning activities.

3.4.6 Third-party peer review. In most cases, plan reviews of proposed projects are performed by the local building department. In certain instances, the local jurisdiction may require a third-party structural peer review to be performed on the project in addition to the plan review performed by their office. These third-party peer reviewers should be licensed design professionals and should have experience performing designs like the one in question. Many times, the permit applicant is required to pay the fee of the third-party peer reviewer in addition to the jurisdiction's plan review and permit fee.

There are several instances within ASCE 41 where an independent peer review should be performed on a project. At a minimum, ASCE 41 recommends that such peer review be performed for all projects for which an NDP is used and for projects that incorporate seismic isolation and energy dissipation techniques.

3.5 *In Situ Load Tests*

Section 302 of the IEBC states that if in situ load testing is required, it shall be conducted in accordance with Section 1708 of the IBC. In situ load testing may be needed quite often when dealing with existing buildings. As an example, when performing an evaluation and rehabilitation design, the ASCE 41 standard requires investigations of in situ conditions when as-built information is not available. For example, consider just the structural steel used within an existing building. Section 9.2.2.1 of ASCE 41 requires that the designer obtain the yield and tensile strength of the base material, the yield and tensile strength of the connection material, and the carbon equivalent for each.

Section 1708 of the IBC allows for in situ load testing if there is reasonable doubt as to the stability or load-carrying capacity of the building or component. Section 1708.2 of the IBC states that this testing must be supervised by a registered design professional and provides two separate procedures that can be followed. The first procedure allows the load test procedure used to follow guidelines outlined in a referenced material standard. The second procedure is to follow a test procedure that is developed by a registered design professional and that meets the minimum criteria outlined in Section 1708.2.2 of the IBC.

3.6 *Accessibility*

In the United States, there are more than 56 million Americans with disabilities, or roughly 19 percent of the population (ANSI, 2015). The odds are that most Americans will experience a disability at some point in their lifetime, whether it is a temporary or permanent condition. Being "disabled" simply means that one has either a physical or mental condition that would limit movement, senses, or activities.

The Americans with Disabilities Act (ADA) is the federal statute that prohibits discrimination against people with disabilities. Title III of the ADA requires that businesses that are generally open to the public, as well as commercial facilities, provide accommodations for those with disabilities, while Title II addresses accommodations for state and local government facilities. These accommodations require that defined portions of buildings be "accessible" for persons with disabilities. The Merriam-Webster dictionary defines accessible as "easily used or accessed by people with disabilities." Wikipedia further states that "accessibility can be viewed as the ability to access and benefit from some system or entity."

Since the ADA has gone into effect, more persons with disabilities are in the workforce today than ever before (ANSI, 2015). The following subsections provide some background in relation to how both the current federal and building code accessibility requirements have come to be, in addition to highlighting the specific IEBC accessibility provisions.

3.6.1 Background. The fight for accessibility rights in the United States has a long and storied history. The following highlights some of the significant milestones that have helped develop today's accessibility provisions for buildings and other facilities (WBDG Accessible Committee, 2017):

- *1961—American National Standard Institute (ANSI) A117.1, Accessible and Usable Buildings and Facilities:* This was the first nationally recognized accessibility design standard. It was originally developed by the University of Illinois and was funded by the Easter Seals Society. It was used by many private-sector businesses in addition to being referenced by several states.
- *1964—Civil Rights Act:* This act made racial discrimination illegal, required employers to provide equal employment opportunities, and required uniform standards for establishing the right to vote.

- *1968—Architectural Barriers Act (ABA):* This was the first law to address accessibility in buildings. It addresses all buildings that are designed, constructed, altered, or leased with federal funds and requires them to be accessible to persons with disabilities.
- *1973—Rehabilitation Act:* This act prohibits discrimination on the basis of disability in programs conducted by federal agencies or in programs receiving federal financial assistance. Section 502 of this act created the Architectural and Transportation Barriers Compliance Board that was later renamed the U.S. Access Board.
- *1984—Uniform Federal Accessibility Standards (UFAS):* This contains accessibility scoping and technical requirements for the implementation of the ABA.
- *1986—ANSI A117.1:* The ANSI standard was updated by the Council of American Building Officials (CABO). The most significant change included the removal of the scoping provisions. Starting in 1986, A117.1 simply provides the technical requirements for accessibility. Standard A117.1 continues to be published by the ICC, its latest edition being ICC A117.1-2017.
- *1988—Fair Housing Amendments Act:* The Fair Housing Act of 1968 prohibited discrimination based on race, religion, sex, or national origin. In 1988 this act was expanded to include persons with disabilities.
- *1990—Americans with Disabilities Act (ADA):* In 1990 the ADA law was signed by President George Bush. It prohibits discrimination based on disability in places of public accommodation, employment, transportation, government services, and telecommunications.
- *1991—Fair Housing Accessibility Guidelines (FHAG):* The U.S. Department of Fair Housing and Urban Development released the FHAG standards to assist builders in complying with the accessibility requirements of the Fair Housing Act.
- *1991—ADA Accessibility Guidelines (ADAAG):* The U.S. Access Board published the ADAAG to explain how buildings can be designed to conform with the ADA. The standard provides both the scoping and technical requirements for new construction and alterations. Subsequent amendments to ADAAG added requirements for transportation facilities, state and local governments, play areas, and recreational facilities.
- *1998—ICC A117.1:* Significant changes to the standard were made to make it easier to use. Figures were placed next to the corresponding text, and Chapters 5 through 10 were added to more appropriately divide up the topics. Starting in the 1998 edition, the A117.1 standard is now overseen by the ICC. The most current version of the standard is the 2017 edition, but the 2018 IBC still references the 2009 edition.
- *2004—ADA/ABA Guidelines:* Updated accessibility guidelines in relation to both the ADA and ABA. This was published by the U.S. Access Board on July 23, 2004, and covers both new construction and alterations to existing buildings. Because of this revision, the guidelines for the ADA and ABA are consolidated into one *Code of Federal Regulations.* Multiple updates have subsequently been made to the standard, with the most recent occurring on May 7, 2014.

As highlighted above, today two standards are used to enforce accessibility requirements in buildings throughout the United States. Perhaps the most well-known is the ADA/ABA Guidelines, which are based on two separate civil rights laws and are not associated with a building code. Whether or not there is construction in a facility, the ADA/ABA Guidelines must be complied with (Rhoads, 2016). The ADA/ABA Guidelines are established by the U.S. Access Board and are enforced by the U.S. Department of Justice, Department of Transportation, Department of Defense, General Services Administration, Department of Housing and Urban Development, and the Postal Service.

The second standard is ICC A117.1 and is enforced by the local code official rather than the federal government. While the ADA/ABA Guidelines provide the scoping and technical guidelines, the ICC A117.1 standard simply provides the technical accessibility requirements, and Chapter 11 of the 2018 IBC provides the scoping provisions. As the IEBC is a companion code to the IBC, the remaining portion of this discussion will focus on the ICC A117.1 standard rather than the ADA/ABA Guidelines.

By following the requirements of IBC Chapter 11, in addition to the ICC A117.1 provisions, the design professional can ensure that sites, facilities, buildings, and elements are accessible to and usable by persons with disabilities. When properly followed, disabled persons will be able to independently get to; enter; and use a site, facility, building, or element.

3.6.2 Existing buildings. Section 301.5 of the IEBC states that existing buildings must comply with the accessibility provisions of the 2009 edition of ICC A117.1. This is worth repeating. *All* existing buildings must comply with the current ICC A117.1 accessibility requirements. While this is specified in the code, the code official cannot simply go to an existing building and start requiring that it be updated to meet the current accessibility requirements. There must first be a trigger for the code official to enforce the accessibility provisions. The IEBC highlights specific triggers for existing buildings when they are undergoing maintenance, changes of occupancy, additions, or alterations. It also includes specific provisions for designated historic structures. The following sections highlight the accessibility triggers when work is being performed in existing buildings. The code official should take every opportunity to ensure that minimum accessibility provisions are provided when one of these triggers is met.

3.6.3 Technical infeasibility. While not allowed in new construction, with existing buildings they can sometimes be exempt from a specific accessibility requirement when it is deemed that the requirement is "technically infeasible." As defined in IEBC 202, technical infeasibility is when providing accessibility is not likely due to structural constraints or other physical or site constraints. Where technical infeasibility exists, the accessibility requirements must be provided to the fullest extent possible.

Figure 3.18 provides an example of when technical infeasibility might occur. In this figure, the highlighted bathroom is required to be accessible, but structural load-bearing walls exist on

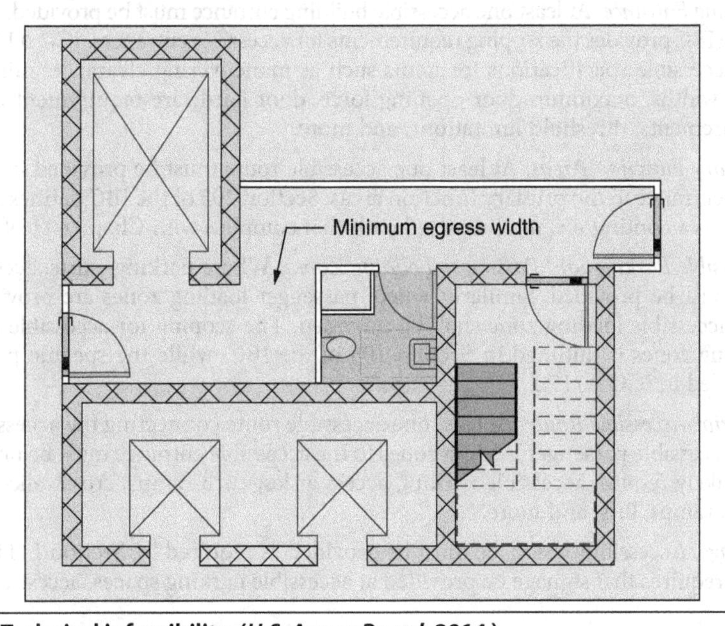

Figure 3.18 Technical infeasibility. (*U.S. Access Board, 2014.*)

two sides while a corridor exists on two sides. Removing load-bearing elements is technically infeasible, and in this instance reducing the corridor width is also technically infeasible, as it is part of the means of egress. In this instance, the bathroom would not have to meet the accessibility requirements, but all other portions of the accessible route must be met.

The U.S. Access Board highlights the following examples to describe when technical infeasibility could occur (U.S. Access Board, 2014):

- *Code Constraints:* Anytime the building, plumbing, life safety, or other code provisions would need to be sacrificed. An example would be combining two water closets to create one accessible stall, yet this would violate the required number of fixtures mandated by the plumbing code.
- *Site Constraints:* Such as meeting the minimum slope requirements on an existing developed site having steep terrain and regrading or other design solutions are not feasible.
- *Structural Constraints:* Work that would impact load-bearing walls and other essential components of the structural frame, including structural reinforcement of the floor slab.

3.6.4 Primary function area. Section 202 of the IEBC defines the primary function area as the major activity for which the facility was intended. Examples provided include the customer service lobby of a bank, the dining area of a cafeteria, meeting rooms within a conference center or offices, and other work areas where activities are carried out. Examples of portions of a building that are not considered to be primary function areas include closets, bathrooms, mechanical rooms, boiler rooms, employee lounges, locker rooms, entrances, and corridors.

3.6.5 Change of occupancy. In general, existing buildings are required to comply with the 2009 ICC A117.1 provisions as required by Section 301.5 of the IEBC; however, Section 305.4 of the IEBC provides the specific scoping provisions for buildings that undergo a change of occupancy. When only a portion of the building is undergoing a change of occupancy, that portion is required to comply with the requirements outlined in Secs. 3.6.7 through 3.6.9.

When a complete change of occupancy occurs, each of the following accessible features must be provided. The requirements for each of those features are outlined in the ICC A117.1 standard.

- *Building Entrance:* At least one accessible building entrance must be provided. Section 1105 of the IBC provides the scoping requirements for accessible entrances. ICC A117.1 provides the accessible specifications for items such as maneuvering clearance, minimum door clear widths, maximum door opening force, door hardware requirements, door closer requirements, threshold limitations, and more.
- *Primary Function Areas:* At least one accessible route must be provided from an accessible entrance to the primary function areas. Section 202 of the IBC defines an accessible route as a continuous, unobstructed path that complies with Chapter 11 of the IBC.
- *Accessible Parking and Passenger Loading Zones:* Where parking exists, accessible parking must be provided. Similarly, where passenger loading zones are provided, at least one accessible loading zone shall be provided. The scoping for accessible parking and loading zones is outlined in Section 106 of the IBC, while the specific provisions are included in ICC A117.1.
- *Exterior Accessible Route:* At least one accessible route connecting the accessible parking and accessible passenger loading zones to the accessible entrance must be provided. This will likely require accessible parking access aisles, curb ramps, crosswalks at vehicular ways, ramps, lifts, and more.
- *Signage:* Accessibility signage must be provided as required by Section 1111 of the IBC. This requires that signage be provided at accessible parking spaces, accessible passenger

loading zones, accessible entrances, accessible checkout aisles, family or assisted-use toilet rooms, accessible dressing or fitting rooms, accessible areas of refuge, exterior areas for assisted rescue and recreational facilities, and lockers that are required to be accessible.

When first reviewing the requirements noted above, the reader may assume that buildings undergoing a change of occupancy must fully conform to the requirements of the 2009 ICC A117.1 and Chapter 11 of the IBC. While many of those requirements must be provided, there are a number of items that are not specifically listed in Section 305.4.2 of the IEBC. An example would be accessible routes from public transportation stops and sidewalks on public streets. In addition, the following two exceptions apply to meeting these accessibility requirements:

- *Technically Infeasible:* If compliance is technically infeasible, the above-noted items are to comply with the maximum extent possible.
- *Type B Dwelling Units:* The accessibility requirements specified above are not required for an accessible route to Type B dwelling units.

3.6.6 Additions. There are two items to be aware of in relation to additions made to existing buildings. One, the addition itself must fully comply with the accessibility provisions outlined in Chapter 11 of the IBC in addition to ICC A117.1. Second, if the addition affects a "primary function area," the requirements highlighted in Sec. 3.6.8 must be met.

3.6.7 Alterations. When an existing building is undergoing an alteration, the altered portions of the building must comply with Chapter 11 of the IBC and ICC A117.1 unless it can be deemed technically infeasible. Where technically infeasible, accessibility shall be provided to the maximum extent possible. The following exceptions apply to portions of existing buildings undergoing alterations:

- *Accessible Routes:* Alterations will not trigger the accessible route requirements unless the alteration affects a primary function area as discussed in Sec. 3.6.8.
- *Accessible Means of Egress:* The accessible means of egress provisions provided in Chapter 10 of the IBC are for new construction only and do not apply to existing facilities.
- *Type A Dwelling Units:* Individually owned Type A dwelling units that are located within a Group R-2 occupancy are allowed to meet the accessibility provisions for Type B dwelling units.
- *Type B Dwelling Units:* If the alteration work is less than 50 percent of the aggregate area of the building, Type B dwelling units outlined in Section 1107 of the IBC would not be required. If more than 50 percent of the aggregate area is altered, the Type B dwelling units would be required.

3.6.8 Primary function areas. Section 3.6.4 defines what a primary function area is. When an alteration affects the primary function area, or the alteration contains a primary function area, an accessible route to the primary function area must be provided. The accessible route must be provided from an accessible parking stall to an accessible entrance, and then from the accessible entrance to the primary function area. The accessible route is also required to have accessible toilet rooms and accessible drinking fountains per Section 305.7 of the IEBC.

While there are numerous exceptions to this requirement, the one that is referenced the most is the "20 Percent Rule." This exception states that the cost of providing the accessible route is not required to exceed 20 percent of the cost of the alterations, which affect the primary function area. To help understand the "20 Percent Rule," consider the following example that was portrayed by Gontram Architecture (2015).

A building owner is remodeling a conference room and a few offices in a three-story office building. The construction budget for the renovation is $100,000. While that is all the owner wants to do, the building has several accessibility deficiencies. While accessible parking is provided, a van-accessible stall is not provided. A ramp is provided to the main entrance, but it does not meet the current ICC A117.1 standards. While the restrooms have grab bars, the stalls are not wide enough. The drinking fountain is not a high-low type of arrangement, and an elevator is not provided.

The owner is obligated to remove those barriers to accessibility up to a cost of 20 percent of the project budget. In this example, 20 percent of the $100,000 budget equates to $20,000 that must be spent toward accessibility improvement. The owner decides to add a van-accessible parking space and re-stripe the parking lot accordingly. In addition, they have chosen to renovate the ramp at the main entrance and to improve the restrooms. The existing drinking fountain will also be replaced with a pair of fountains that meet ICC A117.1. You will remove all barriers up to 20 percent of the original project cost. In performing these accessibility improvements, the owner has met or exceeded the "20 Percent Rule" and therefore an elevator will not need to be installed. While that is the case this time around, if the owner chooses to make renovations in the future they may be required to look at adding an elevator at that time.

The "20 Percent Rule" has been around for some time and was previously included in Chapter 34 of the IBC. Figure 3.19 displays a form that is used by the City of Pittsburgh to help them determine what accessible features will be upgraded as a part of the alteration and to ensure that the 20-percent threshold has been met. Many believe that this exception applies to all accessibility triggers, similar to technical infeasibility, but that is not the case. The "20 Percent Rule" is only for alterations that are made to a primary function area. It cannot be used for changes of occupancy, additions, historic buildings, or relocated structures.

There are also other exceptions in relation to adding the accessible route to a primary function area that has been altered. Those are as follows:

- *Maintenance Alterations:* If the work being performed within the primary function area is limited to windows, hardware, operating controls, electrical outlets, and signs, then accessibility improvements are not required.
- *MEP Alterations:* If the work being performed within the primary function area is limited to mechanical systems, electrical systems, abatement of hazardous materials, or the installation or alteration of fire protection systems, then accessibility improvements are not required.
- *Accessibility Alterations:* If the sole purpose of the work being performed is to increase the accessibility of the facility, then additional alterations will not be required.
- *Type B Dwelling Units:* Altered areas of existing Type B dwelling or sleeping units do not require the addition of an accessible route.

3.6.9 Scoping for alterations. Sections 3.6.5 through 3.6.8 of this chapter outline the accessibility items required for changes of occupancy, additions, and alterations. Section 305.8 of the IEBC provides specific scoping provisions that are in some cases less than what is outlined in Chapter 11 of the IBC. The following outlines the scoping provisions that are specific to alterations within an existing building:

- *Entrances:* If the work involves an existing entrance that is not accessible, it is not required to be made accessible if the building already has an accessible entrance. If that is the case, accessible signage must be provided in accordance with Section 1111 of the IBC.
- *Elevators:* If alterations are made to existing elevators, they shall comply with ASME A17.1 and ICC A117.1. These alterations must also be made to other elevators that are programmed to respond to the same hall call as the altered elevator.
- *Platform Lifts:* While platform lifts are limited in new construction, as outlined in Section 1109.8 of the IBC, the IEBC allows platform lifts to serve as part of the accessible route. If provided, the platform lift must comply with both ICC A117.1 and ASME 18.1.

CITY OF PITTSBURGH
DEPARTMENT OF
PERMITS, LICENSES, AND INSPECTIONS
200 Ross Street, Suite 320, Pittsburgh, PA 15219
phone (412) 255-2175, fax (412) 255-2974

ACCESSIBLE ROUTE COST VERIFICATION FORM
FOR USE WHEN UTILIZING EXCEPTION 1 TO 2012 IBC SECTION3411.7

A. TOTAL COST OF ALTERATION TO PRIMARY FUNCTION AREA:
(To include MEP cost but to exclude costs listed under Item B) $__________

B. COST TO PROVIDE ACCESSIBLE ROUTE:
This includes exterior route from public arrival point and/or from accessible parking spaces (if parking is provided) to accessible entrance.

1. Related to accessible entrance: $__________
2. Related to components of accessible route (Ramps, elevators, platform lifts): $__________
3. Related to accessible parking: $__________
4. Costs associated with toilet room accessible upgrades: $__________
5. Costs associated with accessible drinking fountain: $__________
6. Cost of other accessible upgrades: $__________
   Please explain upgrades:____________________

TOTAL COST OF ACCESSIBLE ROUTE: $__________
This total shall equal or exceed 20% of the cost of item A above.

Responsible Design Professional in Charge:

Name:____________________
Firm/Company:____________________
PA License #:____________________
Phone:____________________
Email/Fax:____________________

Professional Seal:

I certify that the above provided project data is correct.

Signature: ____________________

Figure 3.19 City of Pittsburgh—cost verification form.

- *Stairways and Escalators:* If a stairway or escalator is added within an existing building and significant structural alterations are required for their installation, an accessible route must be provided in accordance with Section 1104.4 of the IBC on the levels served by the added stairway and escalator.
- *Ramps:* If ramps cannot meet the maximum slopes prescribed in Section 1012.2 of the IBC, they can meet the less-stringent requirements listed in Table 305.8.5 of the IEBC. Section 1012.2 of the IEBC requires maximum slopes of 1:12 for any ramps that serve as part of the means of egress. Per Table 305.8.5, the slopes for means of egress ramps associated with existing facilities can be as steep as 1:10.
- *Accessible Dwelling and Sleeping Units:* If the alterations being made affect Groups I-1, I-2, I-3, R-1, R-2, or R-4 dwelling or sleeping units, they must comply with Section 1107 of the IBC. Only those units that are affected as part of the alterations are required to comply.
- *Type A Dwelling and Sleeping Units:* If the alterations being made affect more than 20 Group R-2 dwelling or sleeping units, the quantity of Type A units must comply with Section 1107 of the IBC. Again, only the altered units need to be considered in this determination.
- *Type B Dwelling and Sleeping Units:* Existing Groups I-1, I-2, R-1, R-2, R-3, or R-4 dwelling or sleeping units must meet the Type B accessibility requirements outlined in Section 1107 of the IBC under two conditions:
  - *Condition 1:* It involves an addition or alteration to the space that would add four or more dwelling or sleeping units.
  - *Condition 2:* Where dwelling or sleeping units are being altered and the work area is greater than 50 percent of the aggregate area of the building.
- *Jury Boxes and Witness Stands:* Accessible wheelchair spaces are not required at raised jury boxes and instead can be outside of these spaces when adding a ramp or lift would reduce the width of the means of egress from the space.
- *Toilet Rooms:* When it is technically infeasible to cause existing toilet rooms to meet accessibility requirements, at least one family or assisted-use toilet room shall be provided. This toilet room must be provided on the same floor and in the same area as the existing toilet rooms. Directional accessibility signage shall be provided at the existing toilet rooms, directing the individuals to the family or assisted-use toilet room.
- *Additional Toilet and Bathing Facilities:* If additional toilet fixtures are to be added to assembly and mercantile occupancies or to recreational facilities, at least one family or assisted-use toilet room shall be provided if it would be required for new construction in accordance with Section 1109.2.1 of the IBC.
- *Dressing, Fitting, and Locker Rooms:* If it is technically infeasible to add an accessible dressing, fitting, or locker room at the existing locations, at least one shall be provided on the same level. If separate-sex facilities are provided, at least one accessible space shall be provided for each sex.
- *Fuel Dispensers:* If a replacement fuel dispenser is installed, its operable parts can be a maximum of 54 in. above the vehicular surface. This value is 48 in. under ICC A117.1 for new construction.
- *Thresholds:* The maximum height at thresholds at doorways shall be three-quarters of an inch. These thresholds are also required to have beveled edges on each side.
- *Amusement Rides:* If the structural or operational portions of an existing amusement ride are altered in such a fashion that the ride performance differs from the original design, the amusement ride shall be made to comply with the accessibility provisions outlined for new construction as listed in Section 1110.4.8 of the IBC.

3.6.10 Historic buildings. In general, historic buildings shall comply with the specific accessibility triggers noted above for changes of occupancy, additions, and alterations with two exceptions. The first exception is if the accessible upgrade is technically infeasible. The second exception is if adding the accessible features would threaten the historic significance of the building. If either of those exceptions is triggered, the code official should require that at a minimum the following four accessible features are provided:

- *Site Arrival Points:* At least one accessible route from a site arrival point must be provided to an accessible entrance. This does not necessarily require accessible parking, as an accessible passenger loading zone could be required.
- *Multiple-Level Buildings and Facilities:* An accessible route must be provided from an accessible entrance to all public spaces on that level. Accessibility to spaces on other floors is not required.
- *Entrances:* At least one public entrance must be accessible. If a public entrance cannot be made accessible, another entrance can be used, but it must meet the accessible entrance requirements prescribed in ICC A117.1. Signs complying with Section 1111 of the IBC must be provided at the public entrance and at the accessible entrance. If the accessible entrance is locked during regular business hours, it is required to have a remote monitoring system.
- *Toilet and Bathing Facilities:* If toilet rooms are provided in the facility, at least one accessible family or assisted-use toilet room must be provided. The scoping for this accessible toilet room must comply with Section 1109.2.1 of the IBC.

Figure 3.20 was developed by the U.S. Access Board and displays the minimal accessibility features that are required in historic buildings. Again, historic buildings are required to meet the accessibility requirements highlighted in this chapter for existing buildings unless it is technically infeasible or it will damage the historic fabric of the facility, in which case these lesser requirements must be met.

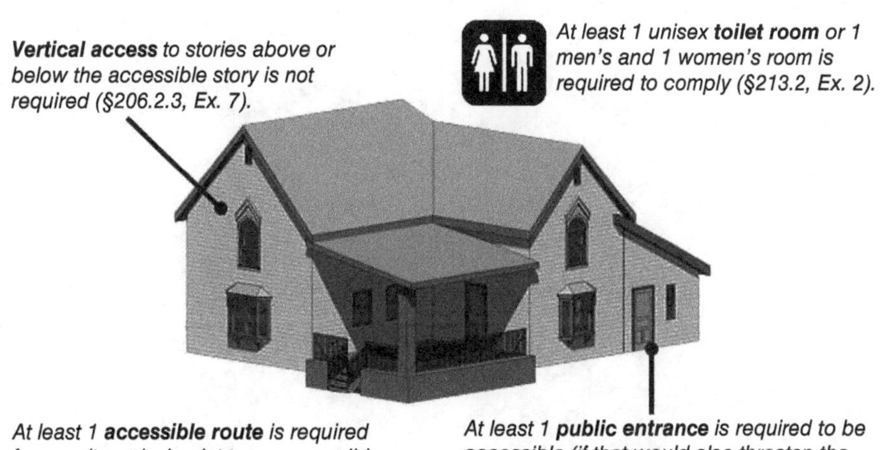

Figure 3.20 U.S. Access Board—exceptions for qualified historic features.

CHAPTER

4

REPAIRS

4.1 *Introduction*

In the 2015 *International Existing Building Code®* (IEBC®), repairs were addressed in Section 404 for the prescriptive compliance method and in Chapter 6 for the work area compliance method. The 2018 IEBC has been reformatted to include all provisions in relation to repairs in Chapter 4. For repairs there is now no distinction between the prescriptive or work area compliance methods. While the scope of the performance compliance method used to also include repairs, it now specifically excludes repairs as all repairs must now comply with the provisions in Chapter 4 of the IEBC.

It is important to note that only historic buildings are not included in the provisions of Chapter 4. For historic buildings, the user must turn to Chapter 12 of the IEBC and review the repair requirements outlined in Section 1202. In general, repairs performed to historic buildings are permitted to be performed using materials similar to the original materials used and can follow the original methods of construction.

Section 401.2 of the IEBC states that repairs can in no way make the building less in compliance with the code than it was before the repair was undertaken. The reader might think this goes without saying, but a quick Google search for "home repairs gone wrong" brings up more than 17 million results. The code official must ensure that repairs are performed correctly and that the building is safe for the occupants.

4.2 *Flood Hazard Areas*

Special requirements for existing buildings located in areas prone to flooding are included throughout the IEBC. In relation to repairs, Chapter 4 of the IEBC discusses existing buildings located within flood hazard areas in both Sections 401.3 and 405.2.5. It is important that both the design professional and the code official pay special attention to the IEBC requirements for existing buildings located in "flood hazard areas" that are undergoing repairs, alterations, or additions.

A flood hazard area, as defined in Section 202 of the *International Building Code®* (IBC®), is essentially an area on a flood hazard map that has a 1 percent chance of flooding in any given year. The Federal Emergency Management Agency (FEMA) has the same definition but uses the term "special flood hazard areas." In layman's terms, flood hazard areas are areas designated on a flood map that could see flood waters during a 100-year flood event. For insurance carriers this means that a building located within a flood hazard area would have a 26 percent chance of being flooded at least once during a 30-year mortgage period (Wittenberg, 2017). When an existing building is located within a 100-year flood zone and it is undergoing significant repairs, alterations, or an addition, it could trigger measures to protect the building due to potential flooding hazards.

This trigger in the code, requiring existing buildings to be brought up to current flood design standards, is an opportunity to reduce future flood damage. Identifying if an existing building is located within a flood hazard area is therefore an important task of the code official. An important objective of the National Flood Insurance Program (NFIP) is to break the cycle of flood damage. As noted in FEMA P-758, "Many buildings have been flooded, repaired or rebuilt, and flooded again. In some parts of the country, this cycle occurs every couple of years."

The NFIP was first authorized by Congress in 1968 with the primary objective of encouraging states and local governments to recognize and incorporate flood hazards in their land use and development decisions. The NFIP is administered by FEMA and has three main elements (FEMA P-758, 2010):

1. Hazard identification and mapping, in which engineering studies are conducted and flood maps are prepared to delineate areas that are predicted to be subject to flooding under certain conditions.

2. Floodplain management criteria, which establish the minimum requirements for communities to adopt and apply to development within mapped flood hazard areas.
3. Flood insurance, which provides financial protection for property owners to cover flood-related damage to buildings and contents.

The statistics show that structures that conform to the floodplain management requirements of the NFIP experience, on average, 80 percent less damage due to either lesser losses due to flooding or reduced frequency of inundation (FEMA P-758, 2010). While floodplain management is often overseen by a local jurisdiction's planning or zoning department, the code official has a very important role in the process, as they ensure that buildings within the designated floodplain are brought up to current standards.

All code officials should have at their disposal the current Flood Insurance Rate Map (FIRM) for their respective jurisdiction. These maps are developed by FEMA and are used by the NFIP for floodplain management, mitigation, and insurance purposes. FIRMs typically show roads and standard map landmarks, the community's base flood elevations (BFEs), designated flood zones, and flood plain boundaries.

Figure 4.1 displays a snapshot of the FIRM developed by FEMA for the Houston, Texas, area. When reviewing a FIRM it is important to recognize flood zone designations that would

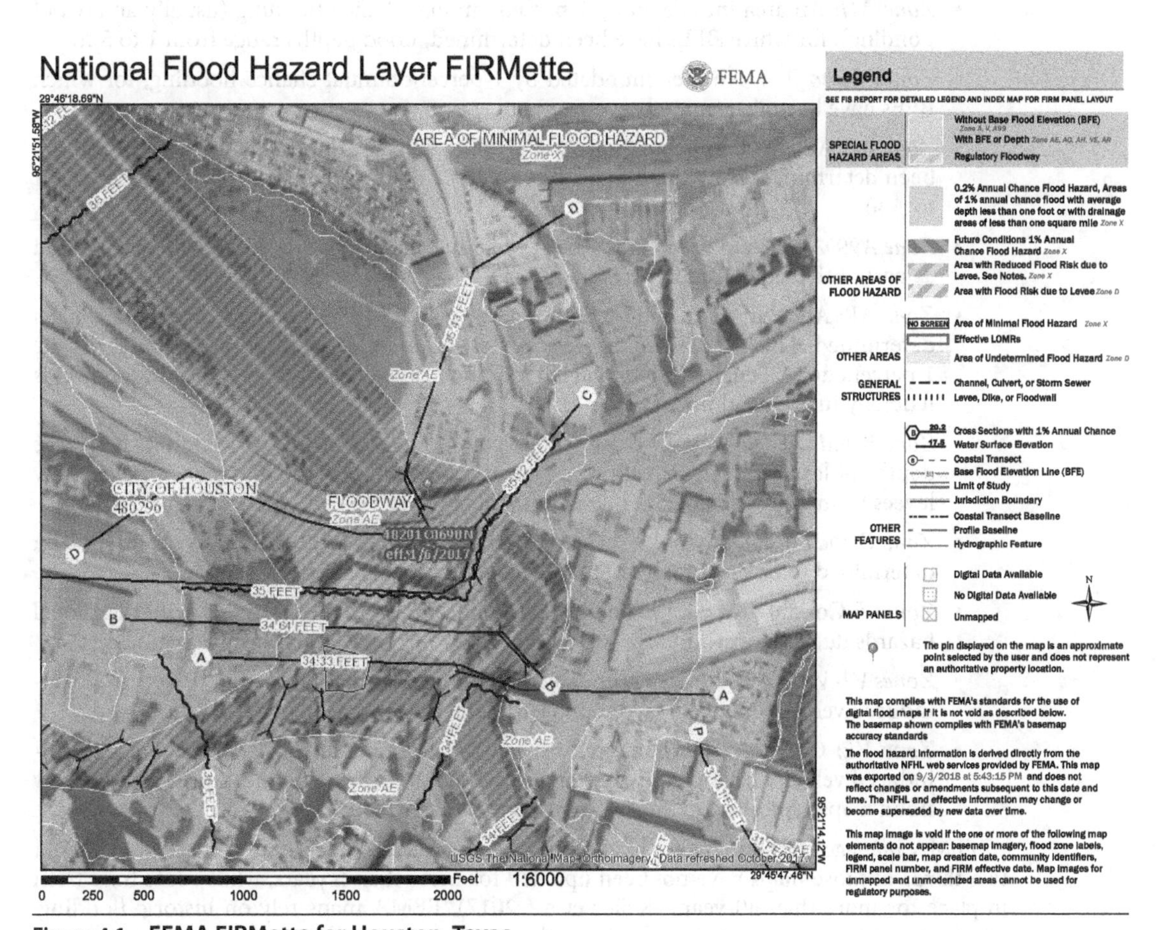

Figure 4.1 FEMA FIRMette for Houston, Texas.

indicate the existing building is located within an area that could be subject to a 100-year flood event. FEMA classifies 100-year flood hazard areas as Zone A, Zone AO, Zone AH, Zones A1 to A30, Zone AE, Zone A99, Zone AR, Zone V, Zone VE, and Zones V1 to V30. That is a lot of zones! This can easily be simplified to state that any flood zone that starts with an "A" or a "V" designates an area that can be inundated during a 100-year flood event. Areas considered to have moderate flooding potential, or that could be inundated during a 500-year flood event, are labeled as either Zone B or a shaded Zone X. Areas of minimal flood hazard are labeled as Zone C or an unshaded Zone X.

In reviewing Fig. 4.1, one can see that the Houston area includes areas designated as floodways, areas listed as 100-year flood hazard areas, areas listed as having a moderate flood hazard, and areas listed as having a minimal flood hazard. The following provides a brief description for each of the FEMA-designated flood zones (FloodMaps.com, n.d.):

- *Zone A:* An area inundated by 1 percent annual chance flooding, for which no BFEs have been determined.
- *Zone AO:* An area inundated by 1 percent annual chance flooding (usually sheet flow on sloping terrain), for which average depths have been determined; flood depths range from 1 to 3 ft.
- *Zone AH:* An area inundated by 1 percent annual chance flooding (usually an area of ponding), for which BFEs have been determined; flood depths range from 1 to 3 ft.
- *Zones A1 to A30:* An area inundated by 1 percent annual chance flooding, for which BFEs have been determined.
- *Zone AE:* An area inundated by 1 percent annual chance flooding, for which BFEs have been determined. New flood maps typically use this designation in place of Zones A1 to A30.
- *Zone A99:* An area inundated by 1 percent annual chance flooding that will be protected as part of the federal flood protection system.
- *Zone AR:* An area inundated by flooding, for which BFEs or average depths have been determined. This is an area that was previously, and will again, be protected from the 1 percent annual chance flood by a federal flood protection system whose restoration is federally funded and underway.
- *Zone B and Shaded Zone X:* Areas of 500-year flood; areas of 100-year flood with average depths of less than 1 ft or with drainage areas less than 1 mi^2; and areas protected by levees from 100-year flood. An area inundated by 0.2 percent annual chance flooding.
- *Zone C and Unshaded Zone X:* Areas determined to be outside the 500-year floodplain; determined to be outside the 1 and 0.2 percent annual chance floodplains.
- *Zone V:* Coastal areas inundated by 1 percent annual chance flooding with additional hazards due to storm-induced waves, for which BFEs have not been determined.
- *Zones V1-V30:* Coastal areas inundated by 1 percent annual chance flooding with storm-induced velocity hazard (wave action), for which BFEs have not been determined.
- *Zone VE:* Coastal areas inundated by 1 percent annual chance flooding with storm-induced velocity hazard (wave action), for which BFEs have not been determined. New flood maps typically use this designation in place of Zones A1 to A30.

FEMA manages flood maps for about 22,000 communities across the United States. About two-thirds of those maps have not been updated for more than 5 years, while some have been in place for more than 40 years (Keller et al., 2017). FEMA maps rely on historic flooding, which can lead to significant inaccuracies such as not accounting for a community's stormwater drainage system, not accounting for new developed areas that introduce large paved areas that drain poorly, and the potential for future storm events to have increased precipitation (Wittenberg, 2017).

While they may not be accurate in many cases, the FEMA flood maps must be relied upon when enforcing the building codes. While every jurisdiction should have access to their own flood maps, the design professional or code official can always go to the FEMA website and review the latest flood zone information (https://hazards-fema.maps.arcgis.com/apps/webappviewer/index.html).

Once the designer, or the code official, has determined that an existing building is within a flood hazard area, they next need to see if the work involved triggers a requirement to update the structure to meet current flood design requirements. When performing repairs, Chapter 4 of the IEBC has two separate and distinct triggers. The first trigger is outlined in Section 401.3 and occurs when the repairs performed constitute "substantial improvement." The second trigger is covered in Section 405.2.5 and is in relation to buildings that have sustained "substantial damage." The following subsections describe the difference between substantial improvement and substantial damage, explain how the designer or code official determines if the improvements or damage are substantial, provide examples of when triggers may be met, and discuss the importance of building inspections during the construction process.

4.2.1 Substantial improvement versus substantial damage. As was defined in Chap. 1 of this book (IEBC Chapter 2, Definitions), substantial improvement can be any repair, alteration, addition, or improvement made to a structure that equals or exceeds 50 percent of the structure's value prior to the improvement. Similarly, substantial damage is defined as any structure that has sustained damage for which the cost of restoration would equal or exceed 50 percent of cost of the undamaged structure.

Note that the key trigger is 50 percent of the value of the structure. Either the owner of the building is performing improvements worth more than 50 percent of the value of the structure, or the damage that has occurred to the existing building requires repairs that will exceed 50 percent of the structure's predamaged value. The 50-percent threshold was established by FEMA and was chosen as a compromise between two extremes. On the one hand, investment in existing buildings located within the flood plain would not occur, and on the other structures could be built within flood plains without any regulation (FEMA P-758, 2010).

Figure 4.2 displays a home that underwent substantial damage as a result of the Boulder County flood in 2013. While it is easy to tell that the repairs in relation to this building will be more than 50 percent of the market value, it is not as easy to determine for the structures shown in Fig. 4.3. This image is of a neighborhood not too far from where the other image is taken. Most of the repairs associated with these buildings would likely be deemed as "less than substantial" but would likely require that the code official carefully consider the repairs that are being made and whether a substantial trigger has been met. The cost of repairing a building that appears to be structurally sound and that has simply been "soaked" during a flood event could include the following items (FEMA 480, 2005):

- Remove and replace all wallboard and insulation
- Tape and paint
- Remove and replace carpeting and vinyl flooring
- Dry the subfloor, replace warped flooring
- Replace cabinets in the kitchen and bathroom
- Replace built-in appliances
- Replace hollow-core interior doors

Figure 4.2 Substantial damage (Jamestown, Colorado). (*Courtesy of FEMA.*)

Figure 4.3 Less-than-substantial damage (Lyon County, Colorado). (*Courtesy of FEMA.*)

- Replace furnace and water heater
- Clean and disinfect duct work
- Repair porch flooring and front steps
- Clean and test plumbing
- Replace outlets and switches, clean and test wiring

4.2.2 Code official determination. It is the responsibility of the code official to determine if substantial improvement is being proposed or if substantial damage has occurred. Design professionals should provide clear information to the code officials to assist them in making this determination. Local jurisdictions should develop clear protocols for how this is to be done and should note all determinations in writing. A good reference for the code official to assist in developing written procedures is FEMA P-758, Substantial Improvement/Substantial Damage Desk Reference.

To make a proper determination, the code official will need to take the following four steps: (1) determine costs, (2) determine market value, (3) make a determination, and (4) require building owners to comply. The key steps are determining the costs associated with improvement or repair, and then comparing that to the market value. To complete step 1, the code official will need to know what costs should be included and which should not. The following are some suggestions that are listed in the FEMA P-758 document noted above:

Included Costs:

- Materials and labor
- Site preparation
- Demolition and construction debris disposal
- Costs associated with other code requirements triggered by the work, such as the accessibility provisions, if applicable
- Costs associated with elevating a structure above the BFE
- Construction management and supervision
- Contractor's overhead and profit
- Sales taxes on materials
- Structural elements
- Interior finishes
- Exterior finishes
- Utility and service equipment

Excluded Costs:

- Clean-up and trash removal
- Temporary stabilization
- Development of construction documents
- Land survey costs
- Permit and plan review fees
- Carpeting
- Outside improvements such as landscaping, irrigation, sidewalks, driveways, fences, yard lights, swimming pools, pool enclosures, and detached accessory structures
- Plug-in appliances

Cost information should be provided to the code official as either an itemized construction cost estimate, a qualified estimate, or it can be based on the ICC® building valuation tables or other cost-estimating manuals. The code official should review the cost information submitted carefully. For buildings that have undergone substantial damage, it may be useful to perform a site visit and to ensure that the proposed work is all that is necessary to restore the building to its predamaged state.

For buildings that have undergone substantial damage, FEMA has developed the Substantial Damage Estimator (SDE) tool. This tool can be used to assess the damage from flood, high wind, wild fires, and earthquakes. The purpose of the tool is to assist code officials in making prompt determinations as to whether or not substantial damage has occurred. This tool can be downloaded for free from the FEMA website.

Figure 4.4 is taken from the FEMA P-758 document and shows the process that the code official must follow to determine the costs of the work. This same process should be followed whether the building is undergoing substantial improvements or substantial damage.

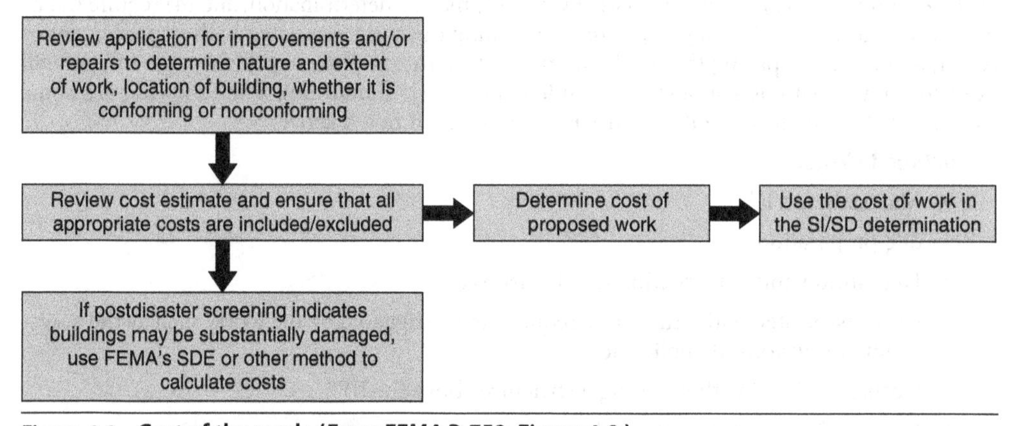

Figure 4.4 Cost of the work. (*From FEMA P-758, Figure 4-2.*)

After determining the costs associated with the improvements and/or repairs, it is important to determine the market value for the structure. When undergoing improvements, this should be the market value before beginning any construction work. If undergoing repairs as a result of substantial damage, the market value should be the value of the structure before the damage occurred.

Sources for obtaining the market value could include an appraisal from a qualified professional, the assessed value used in property tax assessments, estimates of actual cash value (ACV), or qualified estimates based on professional judgment. Whichever method is used, the local jurisdiction should be consistent in what methods they require so as to maintain uniformity. The two most common methods are to use professional property appraisals or property tax assessments. Property tax assessments are not typically allowed for publicly owned or unique and complex buildings. It is important to note that the value of land and accessory structures should not be included in the market value (FEMA, 2017).

Similar to estimating the costs, the FEMA SDE tool can be used to establish the market value of buildings that have undergone substantial damage. Figure 4.5 is taken from the FEMA P-758 document and shows the process the code official should follow to determine the market value of the structure in question.

After establishing the cost associated with the improvements or repairs and ascertaining the market value of the structure, the code official must determine whether the work

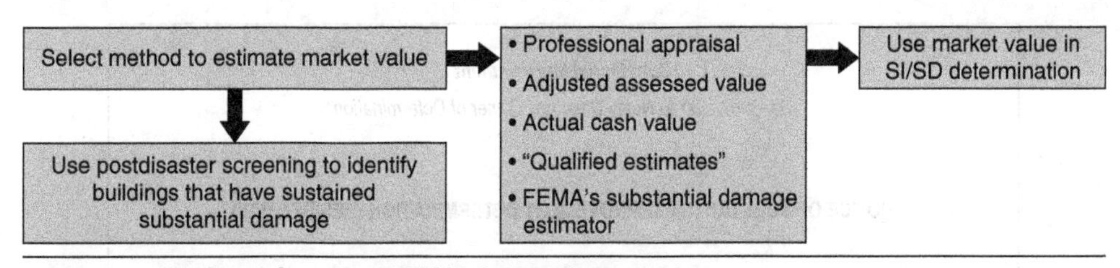

Figure 4.5 **Market value. (*From FEMA P-758, Figure 4-3.*)**

is deemed substantial. To do this, the code official simply divides the established costs by the market value as shown in the equation below. If the value obtained is equal to or greater than 50 percent, the building is undergoing substantial improvements or has sustained substantial damage.

$$\frac{\text{Cost of improvement or cost of repair}}{\text{Market value}} \geq 50\%$$

There may be times that the building owner proposes to do the improvement or repair work in phases. In doing so they will submit separate permit applications for each phase. Each individual phase could be considered "less than substantial," but combined the improvements or repairs being made could constitute a substantial improvement. FEMA recommends that jurisdictions track the changes being made and that if the cumulative result is a substantial improvement, the building should be brought to meet current flood design requirements. It should be clarified that this is a FEMA recommendation and not necessarily a code requirement. Each jurisdiction should determine how they will enforce phased approvals in their region in terms of their flood management practices. While the IEBC does list a 5-year period for substantial structural alterations, no such term has been established for substantial improvements.

After establishing that the work being proposed is equal to or greater than 50 percent of the building's value, the code official should provide an official determination. A notice of the official determination should be sent to the building owner and permit applicant. In the notice the code official should clearly note what the expectations are in terms of bringing the building up to current flood design standards. Figure 4.6 displays a sample notice letter that has been developed by FEMA.

Once the determination has been made, it is the responsibility of the design professional and the code official to ensure that the structure is appropriately brought up to meet current flood design requirements. As stated previously, both Sections 401.3 and 405.2.5 of the IEBC require that the design conform to Section 1612 of the IBC or Section R322 of the *International Residential Code*® (IRC®). The code official should work closely with the owner and design team to assist them in understanding what exactly is required. It would likely be useful to refer to them to guidelines developed by FEMA, such as FEMA P-312, Homeowner's Guide to Retrofitting. This guide provides the building owner with several visuals of how the structure can be retrofitted to comply with current code requirements.

4.2.3 Examples. The purpose of the following examples is to help the reader understand when an improvement to an existing building should be considered substantial. Each of these examples has been taken from the FEMA P-758 document.

Lateral building addition. Figure 4.7 shows two examples of lateral additions attached to existing buildings. Figure 4.7*a* shows that no work was performed in the existing building that would constitute a substantial improvement. With that said, the IEBC requires all new construction to comply with the requirements for new construction, so the addition itself was required to

Substantial Improvement

Sample Letter to Notify Structure Owner of Determination

NOTICE OF SUBSTANTIAL IMPROVEMENT DETERMINATION (RESIDENTIAL)

Dear [name of structure owner]:

The City of Floodville has reviewed your recent application for a permit to [*describe proposed improvement/addition*] for the existing residential structure located at [insert structure address], Floodville, NY 14056.

The Department of Building Inspections has determined that this structure is located within a mapped Special Flood Hazard Area on the Flood Insurance Rate Map (FIRM), Panel 0150, with an effective date of June 19, 2008. As required by our floodplain management ordinance or building code, we have evaluated the proposed work and determined that it constitutes Substantial Improvement of the building. This determination is based on a comparison of the cost estimate of the proposed work to the market value of the building (excluding land value). When the cost of improvements equals or exceeds 50 percent of the market value of the building, the work is considered to be Substantial Improvement under the requirements of the National Flood Insurance Program (NFIP) and the city's Floodplain Management Ordinance dated April 8, 2005.

As a result of this determination, you are required to bring the building into compliance with the flood damage-resistant provisions of the City regulations and/or code [cite pertinent sections].

We would be pleased to meet with you and your designated representative (architect/builder) to discuss the requirements and potential options for bringing the structure into compliance. Several issues must be addressed to achieve compliance. The most significant requirement is that the lowest floor, as defined in the regulations/code, must be elevated to or above the base flood elevation (BFE) [or the elevation specified in the regulations/code] on the FIRM. You may wish to contact your insurance agent to understand how raising the lowest floor higher than the minimum required elevation can reduce NFIP flood insurance premiums.

Please resubmit your permit application along with plans and specifications that incorporate compliance measures. Construction activities that are undertaken without a proper permit are violations and may result in citations, fines, the removal of the non-compliant construction, or other legal action.

Sincerely,

Lisa Donaldson, Chief Inspector
Department of Building Inspections
888-999-0000
lisa.donaldson@floodville.ny.gov

Figure 4.6 Sample substantial improvement notice. (*From FEMA P-758.*)

be elevated above the BFE and designed to meet current flood zone criteria. Figure 4.7*b* shows significant alterations to the existing building along with the construction of the addition. Because of this, the entire structure, including addition, was required to be elevated above the BFE and required to meet current flood design provisions of the code.

Vertical building addition. Figure 4.8 displays a vertical addition that was made to a commercial building. The vertical addition constitutes a substantial improvement and requires that the existing building be brought into compliance with the flood design requirements of the current code. The IEBC refers the designer to Section 1612 of the IBC, which in turn refers the designer to Chapter 5 of ASCE 7 and to ASCE 24, Flood Resistant Design and Construction.

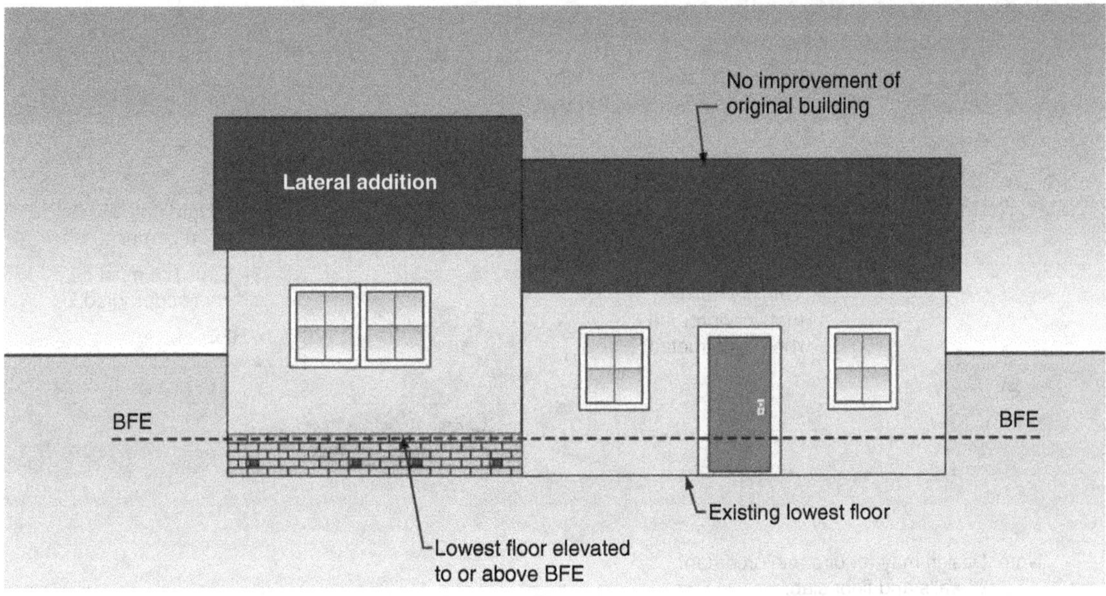

(a)

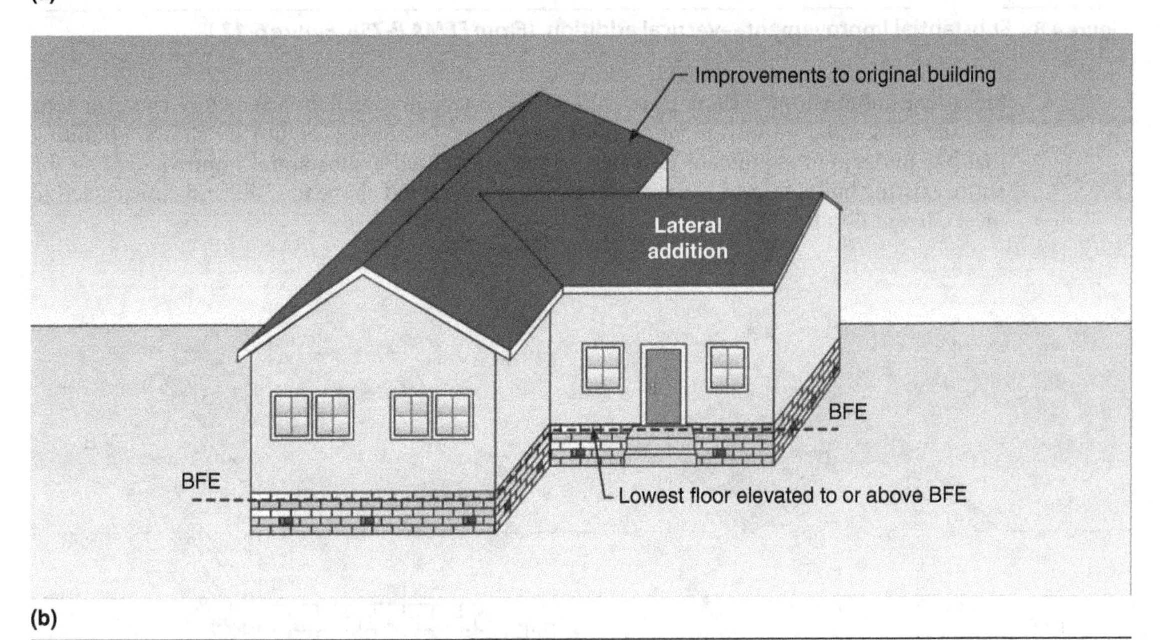

(b)

Figure 4.7 **(*a*) Not a substantial improvement. (*From FEMA P-758, Figure 6-3.*) (*b*) Substantial improvement. (*From FEMA P-758, Figure 6-4.*)**

For nonresidential occupancies, ASCE 24 allows the structure to be "dry floodproofed" up to an elevation above the BFE as shown in Fig. 4.8. Dry floodproofing consists of making the structure watertight by sealing the walls with waterproof coatings, impermeable membranes, or a supplemental layer of concrete or masonry.

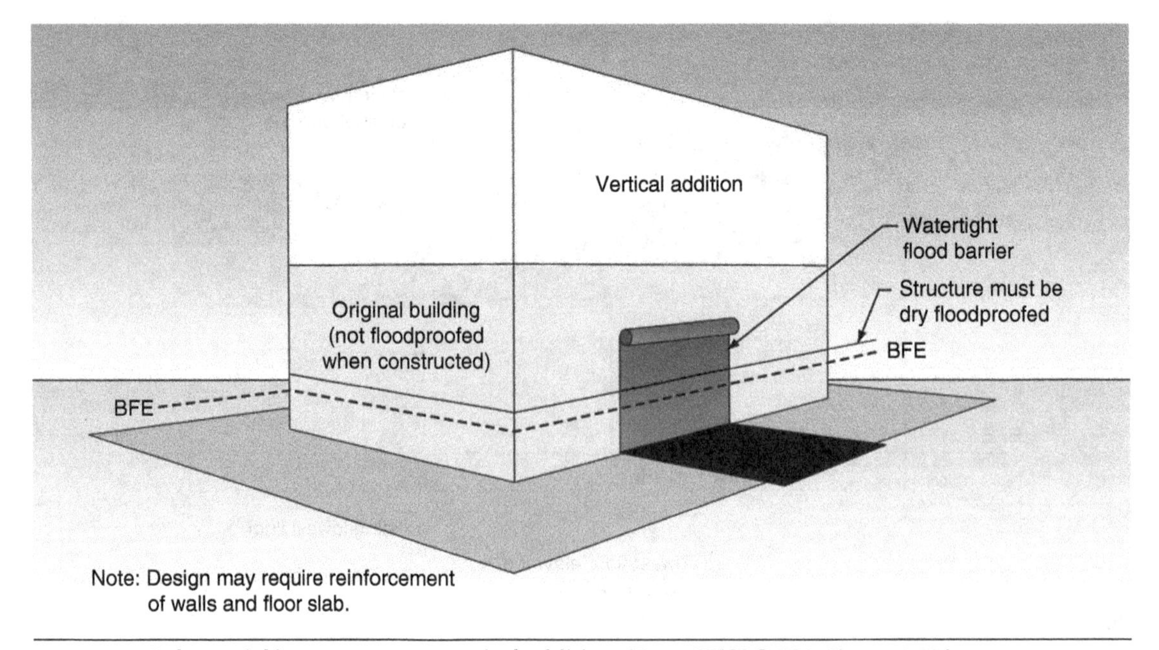

Figure 4.8 Substantial improvement—vertical addition. (*From FEMA P-758, Figure 6-12.*)

Significant alteration. Many times, alterations or repairs made to a structure can also activate the substantial improvement or substantial damage triggers. Figure 4.9 shows a building that has underwent significant alterations that constituted a substantial improvement so the entire existing building, and associated utilities, was elevated above the BFE and constructed to meet current flood design criteria.

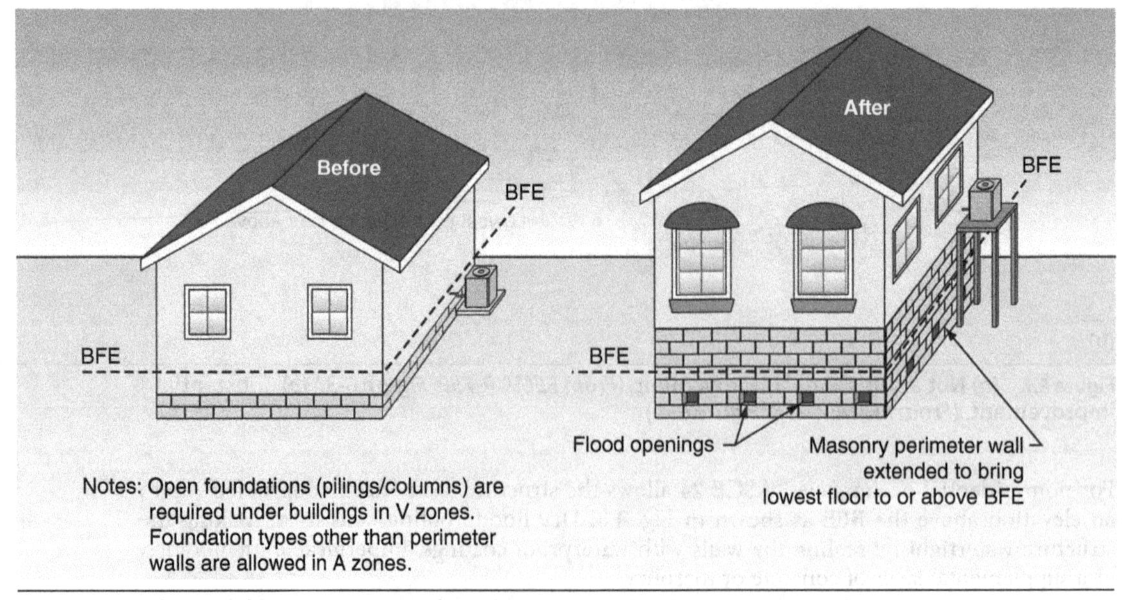

Figure 4.9 Substantial improvement—alteration. (*From FEMA P-758, Figure 6-1.*)

4.2.4 Building inspections. Section 109.3.3 of the IEBC requires that the code official perform an additional inspection when a building located within a flood hazard area undergoes a substantial improvement or incurs substantial damage. In truth, this section does not actually require the code official to perform the inspection, but that the permit applicant submit an elevation certificate to the code official. The purpose of the elevation certificate is to ensure that the lowest occupied floor will be sufficiently above the BFE. This certificate must be sealed by a registered design professional as noted in Section 1612.4 of the IBC. As FEMA oversees the NFIP, they have a formal six-page elevation certificate form (FEMA Form 086-0-33) that should be completed, and this is the certificate that should be provided to the code official.

In addition to requiring the elevation certificate, the code official should be familiar with the flood-resistant-design provisions that are included in the permitted construction documents. As they are performing the other code inspections required by Section 109 of the IEBC, they should be verifying that the flood-resistant construction provisions are being implemented correctly. FEMA recommends that the code officials pay special attention when performing the footing and foundation inspection, the mechanical inspection, enclosure inspection, and final inspection (FEMA P-758, 2010).

4.3 *Safety Glazing*

Section 402.1 of the IEBC requires that all new or replaced glazing that is placed in a hazardous location meet the safety glazing requirements of the IBC or IRC. The only exception to this requirement is that glass block walls and louvered windows (also known as jalousies) can be repaired with like materials. The new safety glazing must meet the impact test provisions of 16 CFR 1201, Safety Standard for Architectural Glazing Material, published by the Consumer Product Safety Commission (CPSC). The impact test requirements must comply with Category II of the CPSC standard. In addition, all safety glazing must be labeled in such a manner that it cannot be removed.

Section 2406 of the IBC defines the "hazardous locations" that would require such safety glazing. The following identifies each of these locations, while similar provisions are provided in Section R308.4 of the IRC.

4.3.1 Glazing in and near doors. All glazing provided in fixed or operable doors must meet the safety glazing provisions. The only exceptions to this include glazing that is small enough that a 3-in. diameter sphere cannot pass through it, if it is decorative glazing (e.g., stained glass), curved glazing that is part of revolving doors, or glazing that is part of commercial refrigerated doors.

In addition, all glazing that is within a 24-in. arc of a fixed or operable door panel and whose bottom edge is within 60 in. of the walking surface must also meet the requirements for safety glazing. Figure 4.10 shows this requirement. The glazing that is shown in the plane of the door as well as the glazing in the two walls perpendicular to the door require safety glazing as they occur within the 24-in. arc. Similar to glazing in doors, glazing adjacent to doors is not required to meet these requirements if a 3-in. sphere cannot pass through or if it consists of decorative glazing. In addition, the glazing panel that is located on the wall perpendicular to the door on the latch side is not required to consist of safety glazing in Group R-3 or Group R-2 occupancies.

4.3.2 Glazing in windows. Glazing that is located within 36 in. of a walking surface shall be safety glazed if the glazing area is greater than 9 ft^2, if the bottom edge of the glazing is within 18 in. of the floor, and if the top edge of the glazing is at least 36 in. above the floor. To get around this requirement, many times a handrail will be placed on the accessible side of the glazing at a height between 34 and 38 in. Another exception to this requirement is if the glazing is decorative, as might be found within many religious buildings.

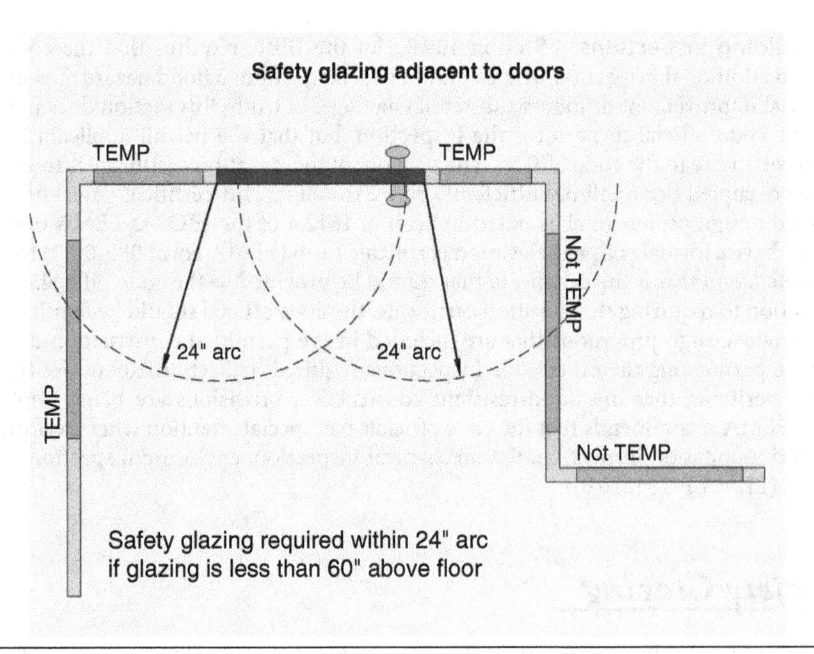

Figure 4.10 Safety glazing adjacent to doors.

4.3.3 Glazing in guards and railings. It is becoming quite common to use glazing materials to meet the guardrail and handrail requirements of the code. In areas where the sights are picturesque, the building owners will often turn to glass guards and railings so as not to diminish their views. The IBC requires that these glass guards and rails consist of safety glazing, regardless of how much glazing area is provided or the height above the walking surface. In addition to meeting the safety glazing requirements, Section 2407 of the IBC provides other material, loading and design requirements for glass guards and rails.

4.3.4 Glazing and wet surfaces. Anytime the bottom edge of glazing is within 60 in. of a standing or walking surface in an area that wet surfaces could occur must consist of safety glazing. The code provides several examples of areas where such wet surfaces could occur, including hot tubs, spas, whirlpools, saunas, steam rooms, bathtubs, showers, and swimming pools. The only exception to this requirement is if the glazing is located at least 60 in. horizontally away from the edge of a bathtub, hot tub, spa, whirlpool, or swimming pool.

4.3.5 Glazing near stairs and ramps. In many cases the glazing that is provided in, or near, stairwells must consist of safety glazing. As such, the code official should pay special attention to glazing provided in these areas. This only applies to glazing along the stair, ramp, or landings whose bottom edge is within 60 in. of the walking surface. In addition, this applies to any glazing that is provided within a 60-in. arc of the bottom stair landing. Figure 4.11 shows the only way out of meeting this requirement. This exception requires a guard to be provided, which separates the walking surface from the glazing in question by a minimum of 18 in.

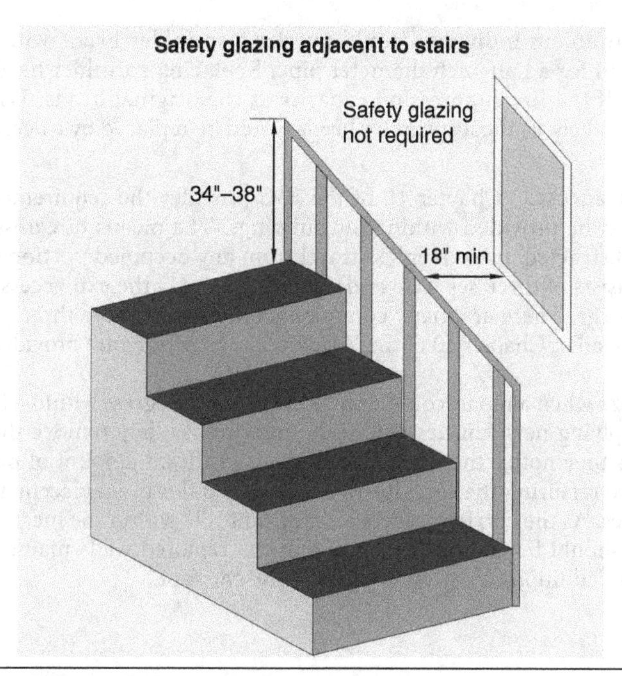

Figure 4.11 Safety glazing adjacent to stairs.

4.4 *Fire Protection and Means of Egress*

Both Sections 403.1 and 404.1 of the IEBC state that any repairs made cannot reduce the level of fire protection or means of egress provided within the building in question. It would seem that this requirement is already included in the IEBC, as Section 401.2 states that the building cannot be any less compliant "…than it was before the repair was undertaken." The reason for the restatement is simply due to the importance of fire protection systems and the means of egress. Both are key life safety components of a building and should never be reduced as part of the work.

4.4.1 Fire protection. Chapter 9 of the IBC discusses fire protection systems that are often provided within a building. Some of the items include fire sprinklers, fire alarms, fire extinguishers, standpipe systems, smoke and carbon monoxide detection, smoke control systems, and much more. The IRC requires significantly less in relation to fire protection, but Sections R313 through R315 address the requirements for fire sprinklers, smoke alarms, and carbon monoxide alarms.

To understand when a repair could reduce the fire protection within an existing building, consider a warehouse building that is fire sprinklered. The building is more than 50 years old, and NFPA® 25 *(Standard for the Inspection, Testing, and Maintenance of Water-Based Fire Protection Systems)* requires that the sprinkler heads be replaced after 50 years of service. The existing sprinkler piping accommodates heads with a half-inch thread diameter. While most standard upright heads that are attached to half-inch pipe have a sprinkler discharge rate (K) of 5.6 L/min, an individual could purchase sprinkler heads with a discharge rate as low as 2.8 L/min for a half-inch diameter pipe. Replacing sprinkler heads in this fashion would provide half the fire suppression capacity as the original heads. With that said, this example is not too likely, as most sprinkler heads would be replaced by a licensed fire sprinkler contractor.

4.4.2 Means of egress. Chapter 10 of the IBC provides the requirements of the means of egress that must be provided within new buildings. The means of egress provides a continuous, and unobstructed, path of egress travel from any occupied portion of a building to a public way. It consists of three separate and distinct parts: (1) the exit access, (2) the exit, and (3) the exit discharge. There are many components to each of these three parts, all of which are discussed in detail in Chapter 10 of the IBC. Similar provisions are provided in Section R311 of the IRC.

To help visualize when a repair could reduce the means of egress within a building, consider a hotel that is applying new finishes within the interior. As they remove the wallpaper from the corridor walls they notice that there had been a significant amount of moisture intrusion, so they commence repairing the corridor walls. This includes the replacement of wallboard in addition to finishes. As the corridor serves a significant role within the means of egress system, the code official should be careful to ensure that the repaired walls maintain the same fire-resistance rating in addition to any openings into the corridor.

4.5 *Structural*

The Merriam-Webster Dictionary defines "repair" as "to restore by replacing a part or putting together what is torn or broken." The key word to focus on in this definition is "restore." Restoration implies that one is bringing something back to its original state. When dealing with the repair of damaged structural elements, that is the default requirement for repairs performed under the provisions of the IEBC. Section 405.2.1 of the IEBC states that structural repairs should restore the element to its "predamage condition."

There are two instances in Chapter 4 of the IEBC that require additional work to be done beyond simply restoring damaged structural members. The first instance is when substantial damage has occurred. This was already discussed in Section 4.2.1 and requires the entire structure to meet the IBC or IRC requirements for flood-resistant construction. The second instance is when substantial structural damage (SSD) has occurred. This will be defined in a moment, but when SSD has occurred, it requires an evaluation of structural elements beyond just those that have been damaged. It could be possible that both substantial damage and SSD have occurred to the same building, in which case both IEBC requirements would need to be met (FEMA, April 2017).

Repairs to structural elements can typically be defined as "less than substantial" or "substantial structural damage." The following subsections will discuss each of these scenarios separately. Upon reading these sections, the reader should have a clear understanding of when a damaged structural element can simply be restored to its predamage condition, or when the damage is significant enough that it requires additional work to be done.

4.5.1 Substantial structural damage. SSD can be caused in a myriad of ways. It could be due to natural events such as tornados, fires, earthquakes, flooding, tsunamis, and more. It could be due to poor construction or expansive, collapsible, or liquefiable soils at the site. Many times, it could also occur simply because the owner has not provided sufficient maintenance of the building or particular components of the structure. Regardless of what the cause of the structural damage is, both the code official and design professional must have a clear understanding of when

SSD is triggered. Section 202 of the IEBC provides a three-part definition for SSD. This definition is very important to understand so the exact code language is provided below:

A condition where any of the following apply:

1. *The vertical elements of the lateral force-resisting system have suffered damage such that the lateral load carrying capacity of any story in any horizontal direction has been reduced by more than 33 percent from its predamage condition.*
2. *The capacity of any vertical component carrying gravity load, or any group of such components, that has a tributary area more than 30 percent of the total area of the structure's floor(s) and roof(s) has been reduced more than 20 percent from its predamage condition, and the remaining capacity of such affected elements, with respect to all dead and live loads, is less than 75 percent of that required by the (IBC) for new buildings of similar structure, purpose and location.*
3. *The capacity of any structural component carrying snow load, or any group of such components, that supports more than 30 percent of the roof area of similar construction has been reduced more than 20 percent from its predamage condition, and the remaining capacity with respect to dead, live and snow loads is less than 75 percent of that required by the (IBC) for new buildings of similar structure, purpose and location.*

The first condition provided in the definition addresses SSD in relation to the building's lateral elements. The second and third conditions both address SSD as it relates to the building's gravity elements. Before discussing the specific lateral and vertical requirement of the IEBC, it is important to discuss how an SSD determination is done. To make the determination, the design professional or code official should have a basic understanding of the building's structural system as well as the extent of the damage that has occurred (FEMA, April 2017). Is the building in question considered a bearing wall system, building frame system, moment frame system, or some other system described in Table 12.2-1 of ASCE 7-16? (See Section 3.3.3 for more information.)

Once the structural system has been determined, the SSD determination will require that the code official review what has been provided by the design professional. For damage that has occurred to lateral elements, the design professional should look at each story, in each direction, and assess the amount that the lateral capacity has been reduced. In relation to the gravity elements, the design professional should check the capacity of each damaged element, or group of elements, and determine how much they have been reduced.

The specific requirements for SSD to lateral or vertical structural elements vary in Chapter 4 of the IEBC. As such, the following provides specific information in relation to each of these separate conditions.

SSD—lateral. The SSD trigger for lateral portions of the building does not occur until a vertical element of the lateral force-resisting system has been damaged by more than 33 percent. That is a very substantial amount. Anything damaged less than that can simply be restored to its predamage condition using like materials and methods.

Figure 4.12 shows an example of when SSD has occurred to vertical components of the lateral force-resisting system. In this image, the masonry foundation was severely damaged during the 2014 South Napa California earthquake. The earthquake occurred on August 24, 2014 and measured 6.0 on the moment magnitude scale. As shown in the figure, the foundation has been reduced in capacity by more than 33 percent.

If SSD has occurred, Section 405.2.3 of the IEBC requires that the building be evaluated and, depending upon the results of the evaluation, damaged elements can be repaired or at times

Figure 4.12 Foundation damage, Napa, California. (*Courtesy of FEMA.*)

may require the building to not only be repaired but retrofitted. The amount of work that will be required is totally dependent upon the results of the evaluation that is provided.

Section 405.2.3.1 of the IEBC details what is specifically required in terms of the evaluation. This section explicitly requires that the evaluation be performed by a registered design professional, and most states' licensure laws would require this to be a licensed professional engineer or licensed structural engineer. This evaluation must be provided to the code official, and it is often recommended that the owner, SER, and code official sit down together to discuss the findings of the report and to determine the next steps that will be required.

As part of the evaluation, the SER shall determine whether, if repaired to its predamage state, the existing building would comply with the structural provisions of the IBC. The one caveat to this is that the seismic portion of the evaluation can use the "reduced" level discussed in Sec. 3.3 of Chap. 3. The full IBC live loads, snow loads, and wind loads must still be considered in the analysis.

If the analysis shows that the existing building would be compliant if simply repaired, the damaged structural elements can simply be brought back to their predamage condition. If the evaluation shows that the building would not be compliant if simply repaired, a retrofit design must be provided, and appropriate construction documents and structural calculations submitted for a building permit. The retrofit design must comply with the following:

1. The live loads specified in Section 1607 of the IBC.
2. The snow loads required by Section 1608 of the IBC.

3. The wind loads required by the code in effect at the time of original construction. If the SSD that occurred was due to wind, the current IBC wind requirements must be met.
4. The seismic design requirement of the code in effect at the time of original construction or the reduced IBC seismic loads, whichever is greater.

SSD—gravity. The SSD gravity trigger is a bit more difficult to comprehend. While it is relatively easy to determine if a vertical component of the lateral-resisting system has been reduced by 33 percent, the SSD gravity trigger involves a three-step process. First, the design professional must determine if the damaged vertical element (e.g., wall or column) is supporting up to 30 percent of the roof or floor area. That is a huge portion of the building and is typically not the case. The second step is to determine if that vertical component has been reduced in capacity by 20 percent or more. If both conditions exist, the design professional must check the existing load-carrying capacity of the element as required by the current code. It is only considered SSD if the capacity, when compared to the current code, is less than 75 percent of what is required. That is a very complicated process to follow just to determine if SSD has occurred.

When it is not clear if SSD has occurred, either to the lateral or gravity structural members, the code official is able to require thorough documentation, testing, analysis, or even third-party peer reviews to help them in the determination. While the description provided above for the gravity SSD trigger is complicated, often the damage sustained is so minimal, or so severe, that the design professional or code official can make the SSD determination without any form of analysis (FEMA, April 2017).

Figure 4.13 provides an example of when the SSD to gravity elements would be triggered. This image is of a home in Valdosta, Georgia, and shows damage from a falling tree that occurred during Hurricane Hermine in 2016. The tree obviously damaged more than 30 percent of the upper roof area, and the capacity of the elements damaged has been reduced by well over 20 percent. In this instance an analysis would not be required to show that SSD has occurred.

Once the SSD gravity determination has been made, Section 405.2.4 of the IEBC states that the gravity elements that have sustained damage are to be rehabilitated such that they comply with the IBC in relation to dead and live loads. Snow loads only need to be considered in the analysis if the damage that occurred was due to snow load effects.

In addition, Section 405.2.4 requires that undamaged structural elements that are supporting the repaired elements be checked to meet the IBC gravity load requirements. If they do not comply they should be retrofitted along with the damaged members.

At times, it may not be apparent that the lateral-force-resisting system of a building has been damaged in a wind or seismic event, but gravity members did incur SSD in those events. This could happen when a building has not been detailed appropriately, and the lateral loads are not transferred to the lateral-resisting system and instead are resisted by gravity elements not designed for that purpose. When SSD occurs to gravity members during a seismic or wind event, Section 405.2.4.1 of the IEBC requires that an evaluation be performed in accordance with Section 405.2.3.1. If the evaluation shows the building is not in compliance, the building will need to be retrofitted as discussed previously.

4.5.2 Disproportionate earthquake damage. Section 405.2.2 of the IEBC requires that buildings that have sustained disproportionate earthquake damage and are assigned to Seismic Design Categories D to F shall be subject to the requirements of the SSD lateral requirements discussed above. This requirement is new in the 2018 IEBC, and there is not much available on the Internet in relation to this subject. How does one determine if disproportionate earthquake damage has occurred?

Figure 4.13 Tree on house, Hurricane Hermine, Valdosta, Georgia. (*Wikimedia, 2016b.*)

Section 202 of the IEBC defines "disproportionate earthquake damage" as earthquake-related damage that includes both of the following:

1. The 0.3-second spectral acceleration at the building site as estimated by the United States Geological Survey (USGS) for the earthquake in question is less than 40 percent of the mapped spectral acceleration parameter S_S.
2. The vertical elements of the lateral force-resisting system have suffered damage such that the lateral load-carrying capacity of any story in any horizontal direction has been reduced by more than 10 percent from its predamage condition.

Each of the items above needs to be discussed one at a time. To understand the first condition, it is important to know what is meant by spectral acceleration (SA) and S_S, as referenced in the definition above. SA is a term used throughout the IBC, and its referenced standards, to define the earthquake motions that a building would need to be designed to resist. SA is a unit measured in "g" (e.g., acceleration due to earth's gravity). The peak ground acceleration (PGA) is a term used to define the motion experienced by a particle on the ground, while the SA represents the motion that is experienced by the building itself.

For design purposes the IBC requires that a building be designed considering the SA at 0.2 s as well as the SA at 1.0 s. The first is known as the short period SA (S_S), while the second is the 1-s SA (S_1). The SER must consider both of these values when determining the controlling seismic forces that the building will need to resist.

Now, returning to the first condition included in the definition. It states that in order to be considered "disproportionate earthquake damage," the 0.3-s SA must be less than 40 percent of the S_S. Other than Section 405.2.2 of the IEBC, the 0.3-s SA is not discussed within any of the International codes. As noted previously, the seismic design parameters in the IBC are based solely on the adjusted S_S and S_1 values. So how is the 0.3-s SA information obtained?

To obtain the 0.3-s SA, the design professional must refer to the USGS "ShakeMap" archives for the site in question (https://earthquake.usgs.gov/data/shakemap/). ShakeMaps provide near-real-time maps of ground motions and shaking intensity following significant earthquakes. Figure 4.14 provides an example of a USGS ShakeMap. This particular example is from a seismic event that occurred on January 17, 1994, near Reseda, California. The USGS site provides SA values at 0.3, 1.0, and 3 s. Figure 4.15 provides a zoomed-in view of the 0.3-s SA ShakeMap shown in Fig. 4.14. Assume that the existing building in question is near Gavin Canyon, California. As shown in Fig. 4.15, the 0.3-s SA that should be considered is 1.4 g (e.g., 140% g).

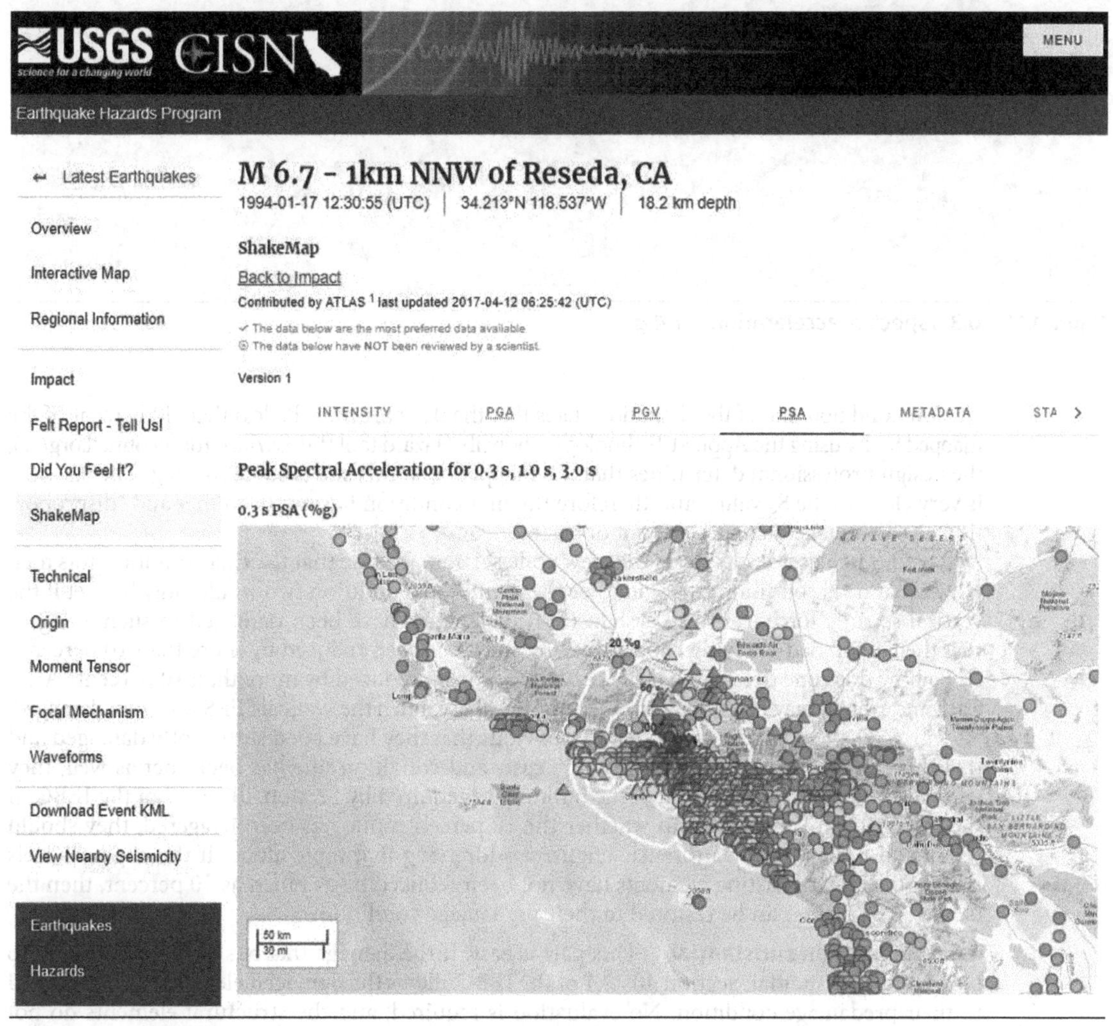

Figure 4.14 Sample USGS ShakeMap, near Reseda, California.

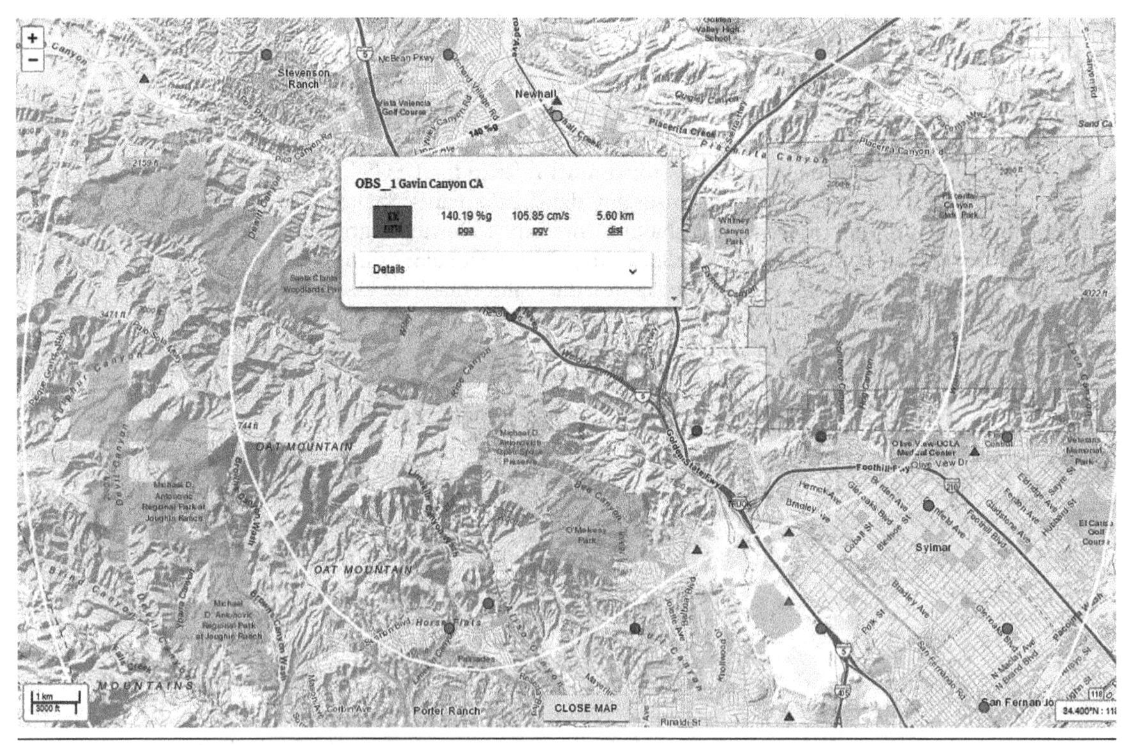

Figure 4.15 **0.3-s spectral acceleration—1.4 g.**

Now condition one of the definition states that the 0.3-s SA must be less than 40 percent of the mapped S_S. By using the Applied Technology Council's Hazard tool (https://hazards.atcouncil.org/#/), the design professional determines that the mapped S_S at this site is equal to 1.5 g. The 0.3-s SA is very close to the S_S value, and therefore the first condition has not been met and "disproportionate earthquake damage" has not occurred.

In order to better understand this new code section, assume that the first condition was met. Now, what does condition two state? Per the definition, condition two would only be met if the vertical seismic force–resisting system (SFRS) elements have been damaged to such a degree that the lateral load-carrying capacity of any story has been reduced by more than 10 percent.

So how does one determine if the capacity has been reduced by more than 10 percent? After a seismic event it may be difficult to determine just how much the vertical SFRS was been damaged. As with SSD, there will be times that it is obvious that they have been sufficiently damaged and that this has been triggered. If that is the case, and condition one has been met as well, they will need to provide an analysis of the building as required by Section 405.2.3.1 of the IEBC. If the code official is unsure as to whether the 10 percent value has been triggered, they should require an evaluation by a forensic engineer addressing that topic alone. If the engineer feels that vertical load-resisting elements have not been reduced by as much as 10 percent, then the damaged elements can be restored to their predamage condition.

4.5.3 Less than substantial. This part is easy. If the damage that has occurred is found to be less than substantial, Section 405.2.1 of the IEBC allows the damaged elements to be restored to their predamage condition. No evaluation is required, and the structural elements do not need to be shown to meet the current IBC or IRC requirements. As discussed in Chap. 3, new and replacement materials should not contain hazardous materials, and the code official still has the right to require any unsafe or dangerous conditions to be remedied.

There is one exception that does not allow less-than-substantial damage to simply be restored to the predamage condition. This is described in Section 405.2.1.1 of the IEBC and occurs when the damage to the structural element is due to snow loads. If this is the case, the damage must be repaired in such a fashion as to comply with the current snow load provisions in Section 1608 of the IBC.

Figure 4.16 shows just such an instance. In this case, the canopy in front of Diego's Hair Salon in Washington, D.C., was damaged during a snowstorm in 2010. In this instance it is not adequate to simply reinstall the canopy as it was before. The design professional must provide an analysis to the code official showing how the canopy, and its support structure, can resist the snow loads prescribed in the IBC and also provide detailed construction documents for the installation. For the canopy, the analysis should not just consider a uniform roof snow load but also snow drift loads that will occur (see Sec. 3.3.2 in Chap. 3 for a more detailed discussion of snow drift loads).

It is important to note that the snow loads prescribed in Section 1608 of the IBC are general in nature. Figure 1608.2 of the IBC includes several areas that are specified as case study areas. The IBC states that the design professional must refer to site-specific case studies for snow loads in these areas. It is always a good idea for the design professional to contact the local jurisdiction to determine what the appropriate snow load is for the project.

Consider the example shown in Fig. 4.16. The IBC prescribes a ground snow load of 25 lb/ft^2 in Washington, D.C. The construction codes adopted by the District of Columbia require that

Figure 4.16 Property damage at Diego's Hair Salon, Washington, D.C. (*Wikimedia, 2010.*)

a minimum 25 lb/ft^2 snow load be considered in addition to snow drift, or a total of 30 lb/ft^2 roof load, whichever is greater. This is a significant difference from what is prescribed in the IBC. As such, the design professional should always check with the local jurisdiction prior to commencing the design to ensure that appropriate snow loads are being used.

4.6 *Electrical*

Section 406.1 of the IEBC allows existing electrical wiring and equipment that is being repaired to be replaced with like material. While that is the general case, there are five exceptions to the rule. Each of those exceptions is explained below.

4.6.1 Receptacles. All receptacles that are replaced must comply with the requirements of Article 406.4(D) of the *National Electrical Code*® (NEC®). This section requires replacement receptacles to meet the following requirements:

Grounding-type receptacles. Grounding-type receptacles must be used if a grounding means exists in the receptacle enclosure or an equipment grounding conductor is installed. The requirements for grounding-type receptacles are covered in Article 406.0 of the NEC.

Nongrounding-type receptacles. In the instance that an equipment grounding conductor does not exist, Article 406.4(D)(2) of the NEC provides the following three options when replacing a nongrounding-type receptacle:

- Nongrounding-type receptacle can be replaced with another nongrounding-type receptacle, or
- Nongrounding-type receptacle can be replaced with a ground-fault circuit-interrupter (GFCI) receptacle so long as a cover plate is labeled as "no equipment ground," or
- Nongrounding type receptacle can be replaced with a grounding-type where supplied by a GFCI. If such is the case, the grounding-type receptacle or cover plate shall be marked "GFCI protected" and "no equipment ground" must be visible after installation.

Ground-fault circuit-interrupters (GFCIs). Replacement receptacles that occur in locations where the NEC requires GFCI protection shall be replaced with GFCI receptacles. GFCI protection can sense an imbalance in current between the ungrounded (e.g., hot) conductor and the neutral conductor and will trip when there is an imbalance of 6 mA or more. GFCI protection helps to prevent electric shock and electrical fires. Without getting into all the provisions of the NEC, such as distinguishing between commercial and residential requirements and the specific exceptions, GFCI receptacles are typically required in the following locations:

- Bathrooms
- Kitchens
- Rooftops
- Outdoors
- Within 6 ft of the outside edge of sinks
- Indoor wet locations
- Locker rooms and showers
- Garages, services bays, and similar areas

- Crawl spaces
- Unfinished portions of the basement that are not intended to be habitable
- Boathouses and boat hoists
- Electrically heated floors

Arc-fault circuit-interrupter protection (AFCI). When receptacles are replaced in dwelling units or dormitory units and the NEC would require AFCI for new construction, AFCI protection shall be provided. This can be provided by either installing an AFCI-protected receptacle or ensuring that the receptacle is protected by a listed branch circuit-type AFCI. The purpose of AFCI is to shut off the flow of electricity when it detects that an arc is about to occur. Arcs are a sudden surge of power, and often standard branch circuit breakers may not trip before the arc occurs. AFCI can prevent electrical shorts and is an important component to prevent electrical fires. Without going through all the NEC provisions, AFCI is typically required for all 120-V, single-phase, 15- and 20-A branch circuits supplying receptacles installed in the following locations within a dwelling unit or dormitory unit:

- Bedrooms
- Family rooms
- Living rooms
- Hallways
- Closets
- Kitchens
- Bathrooms
- Laundry rooms
- Libraries/dens
- Sunrooms
- Recreation rooms

Tamper-resistant receptacles. If the NEC requires tamper-resistant receptacles for new construction in areas where the existing receptacle is to be replaced, a tamper-resistant receptacle shall be installed. The only exception to this is where a nongrounding receptacle is replaced with a nongrounding receptacle. Tamper-resistant receptacles incorporate an internal mechanism that limits access to energized internal components. Every year an estimated 2400 children are injured by electrical outlet-related incidents (Hubbell, n.d.). Because of this, the NEC requires that tamper-resistant outlets be installed in areas that would allow children to access the receptacle. The NEC requires tamper-resistant receptacles at the following locations:

- Dwelling units (includes wall spaces, small-appliance circuits, countertop space, bathroom areas, outdoors, laundry areas, garage and outbuildings, and hallways)
- Guest rooms and guest suites of hotels and motels
- Child care facilities
- Preschools and elementary education facilities
- Business offices, corridors, waiting rooms and the like in clinics, medical, and dental offices, and outpatient facilities

- Assembly occupancies that include areas awaiting transportation, gymnasiums, skating rinks, and auditoriums
- Dormitories

Weather-resistant receptacles. All receptacles that are replaced in areas where the NEC would require new installation to be weather-resistant shall be replaced with weather-resistant receptacles. These receptacles must be listed, labeled, and often require a protective enclosure. Without covering all of the NEC provisions in detail, including any relevant exceptions, listed weather-resistant receptacles are required in the following locations:

- Damp locations (e.g., outdoor partially protected areas such as under eaves, canopies, porches, etc.)
- Wet locations (e.g., unprotected areas exposed to weather, vehicle washing areas, underground installations, or other areas subject to water or liquid saturation)

4.6.2 Plug fuses. While most buildings constructed since the early 1960s have electrical service panels that use circuit breakers, prior to the 1960s a fuse box was the most common method for providing overcurrent protection to electrical circuits. The fuse box has a series of threaded sockets in which the fuses are screwed into like a light bulb. Like circuit breakers, each circuit in the home is protected by a fuse, and it is important that the correct type and rating are selected. Using the wrong type of fuse could pose a serious fire hazard (Formisano, 2018).

As shown in Fig. 4.17, there are two different types of plug fuses. The fuse shown on the left is known as an Edison base, while the fuse on the right is known as a rejection base. Note that the Edison base fuse has brass threads, while the rejection base has porcelain threads. A rejection base fuse is considered "tamper-proof" and has a specific size of thread for each amperage rating to prevent mismatching the fuses. On the other hand, Edison base fuses of any rating can fit into any Edison socket. If someone places a 20-A Edison base fuse in a 15-A electrical circuit, it creates a condition called "overfusing" and can result in the fuse failing to blow before the circuit overheats.

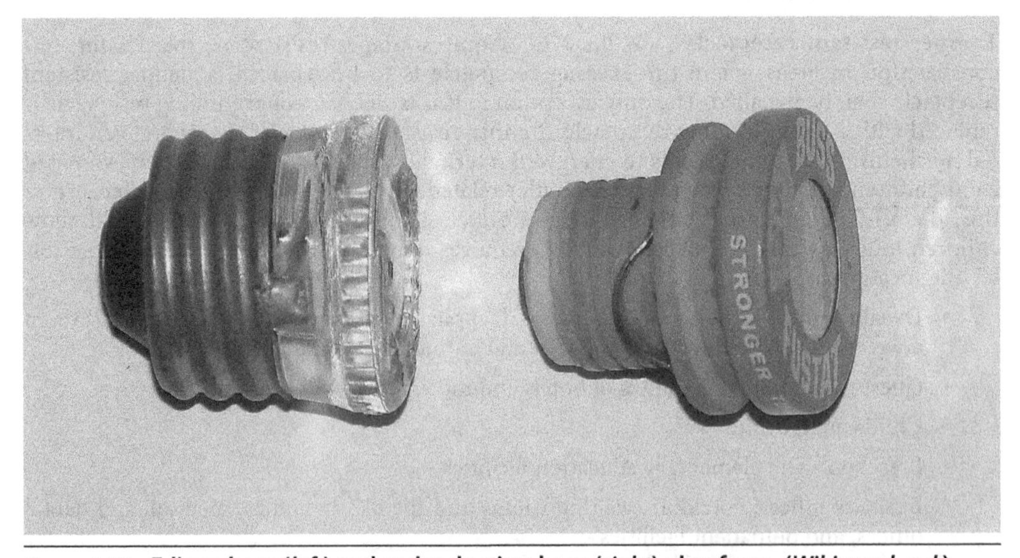

Figure 4.17 Edison-base (*left*) and and reduction-base (*right*) plug fuses. (*Wikiwand, n.d.*)

Section 406.1.2 of the IEBC requires that Edison base fuses be replaced if there is any evidence of overfusing or tampering. When reviewing Fig. 4.17 one can see that the bases for each fuse type are different. For this reason, a socket adapter is required to properly fit a rejection base fuse into an Edison socket. Because the rejection base fuse threads vary by rating, the correct size socket adapter must be selected in each case. Many electricians recommend that all Edison base fuses be replaced with rejection base fuses, regardless of evidence of tampering or overfusing.

4.6.3 Nongrounding-type receptacles. If nongrounding-type receptacles are replaced with grounding-type receptacles, the grounding conductor of the replacement receptacle shall be grounded to any accessible point on the grounding electrode system. This shall be done in accordance with Section 250.130(C) of the NEC.

4.6.4 Group I-2 receptacles. All receptacles located in patient bed areas of Group I-2 occupancies shall be "hospital grade" as required by NFPA 99 and Article 517 of the NEC.

4.6.5 Grounding of appliances. Electrical appliances, such as electrical ranges, cooktops, clothes dryers, and the like, which are installed on an existing branch circuit shall be grounded to the existing grounded circuit conductor in accordance with Section 250.140 of the NEC.

4.7 *Mechanical*

Section 407.1 of the IEBC allows repairs to be made to existing mechanical systems so long as the repairs do not make the building less compliant with the code than it was before the damage occurred. As noted in Section 105.2 of the IEBC, this includes the replacement of any part that would alter the listing or approval of the equipment.

In addition, Section 407.2 of the IEBC allows the installation of mechanical draft systems for manually fired appliances or fireplaces. This requirement is identical to what is specified in Section 804.3.8 of the *International Mechanical Code*® (IMC®). The concerns surrounding mechanical draft systems relate to how the flow of fuel will be addressed if the system malfunctions due to electrical or mechanical failure. To address this, both the IEBC and IMC require the following three conditions to be met if a mechanical draft system is to be installed:

1. The mechanical draft device must be listed in accordance with UL 378, standard for draft equipment, and the installation must comply with the manufacturer's installation instructions.
2. A device must be installed that produces both an audible and visible warning upon failure of the mechanical draft device or loss of electrical power. The device must be equipped with battery backup.
3. A smoke detector equipped with battery backup must be installed in the room of the appliance or fireplace.

It is interesting to note that the requirements of the IEBC and IMC are almost identical to what is specified in the NFPA 211 standard for chimneys, fireplaces, vents, and solid fuel-burning appliances. The one difference between them is that in addition to the three items listed above, the NFPA 211 standard requires a carbon monoxide warning device to be installed. It is not clear if this must be integral with the mechanical draft equipment or if a combination smoke–carbon monoxide detector can be placed in the same room (Enervex, 2011). Regardless, this requirement is not in the IEBC and would not be required unless enforced by the local jurisdiction.

4.8 *Plumbing*

Section 408.1 of the IEBC states that plumbing materials that are prohibited for use in the *International Plumbing Code*® (IPC®) should not be used for repairs. Section 303 of the IPC lists the general limitations for plumbing materials. In essence, all plumbing materials that are used for repairs should conform to the following requirements:

1. All materials shall bear an identification of the manufacturer with any necessary markings to show compliance with applicable standards (e.g., UL listings).
2. All plastic pipe, fitting, and components must be third-party certified and conform to NSF 14.
3. All plumbing products and materials must be listed by a third-party certification agency to show compliance with the IPC-referenced standards.

The only other requirement that the IEBC has in relation to plumbing repairs pertains to water closets (e.g., toilets). Section 408.2 of the IEBC requires that replacement water closets have a maximum water consumption of 1.6 gallons per flush. The only exception to this is to install a blowout design water closet, in which case the maximum water consumption would be 3.5 gallons per flush. These requirements are quite standard in the industry now, and it is quite difficult to purchase a water closet at your local home store that does not meet these limitations. In fact, most water closets sold on the market today have a water consumption of significantly less than 1.6 gallons per flush.

CHAPTER 5

PRESCRIPTIVE METHOD

5.1 *General*

The prescriptive compliance method is perhaps the most used of the three options outlined in the *International Existing Building Code*® (IEBC®). To understand the prescriptive approach, it is helpful to review the definition for "prescribe." The Merriam-Webster Dictionary defines "prescribe" as "to lay down a rule." In essence, the prescriptive approach provides a set of rules based upon the work that is being performed. The work is broken down within Chapter 5 of the IEBC as additions, alterations, fire escapes, windows and emergency escape openings, change of occupancy, and historic structures. The design professional is required to comply with the specific rules outlined within each of those sections, depending upon the actual work being performed.

To understand why the prescriptive approach has been the most used, it is important to look at the history of the *International Building Code*® (IBC®). As noted in Chap. 1 of this book, the first version of the IEBC was published in 2003. Upon publication, it took several years before jurisdictions began adopting the IEBC. Rather, most code officials required all work in relation to existing buildings to conform to the requirements of the IBC. Since the inception of the IBC in 2000, Chapter 34 of the IBC included specific provisions for existing buildings. These original provisions formed the basis for the prescriptive compliance method that was included in the 2003 IEBC.

A significant change was made in Chapter 34 of the 2009 IBC, as it became the first version of the IBC to directly reference the IEBC. While the 2009 IBC referenced the IEBC as an alternate approach to complying with the approach outlined in Chapter 34, the 2015 IBC took another huge step forward by removing Chapter 34 in its entirety. Since the 2015 IBC, existing buildings now are required to comply with the IEBC. Chapter 1 of the IBC directly references the IEBC, similar to the mechanical, plumbing, and other referenced codes. Because the prescriptive approach in the IEBC is based upon the previous IBC Chapter 34 provisions, and these provisions were enforced by jurisdictions throughout the United States for a number of years, most design professionals are more comfortable and familiar with this approach.

The prescriptive approach is the simplest method to follow in terms of code requirements, but may also be the most conservative approach. Many design professionals will select this option for smaller projects, but if significant modifications are being made, they may decide to use the work area method. The performance compliance is not used that often, but typically is only an option for newer existing buildings.

It is interesting to know that the State of Wisconsin, under Section 366.0400 of the Administrative Code, has removed the prescriptive compliance method as part of their adoption of the IEBC. They chose to remove this option as it was deemed to be overly subjective. The following sections outline the specific rules associated with each type of work under the prescriptive compliance method.

5.2 *Additions*

Section 202 of the IEBC defines additions as "an extension or increase in floor area, number of stories, or height of a building or structure." Often building owners feel that additions are simply a horizontal or vertical addition to an existing building. Many times, an addition can occur within an existing space without increasing the height or horizontal dimensions of the building. Consider an existing attic space in a home that is being converted to living space. Another example would be adding a mezzanine level within an existing warehouse building. Whether or not it is a vertical addition, horizontal addition, or an addition within the existing space, the requirements of this section must be met if following the prescriptive compliance path.

It is first important to understand that the addition itself must fully comply with the requirements of the IBC for new construction. This is worth repeating: the addition must comply with the IBC in its entirety. The addition together with the existing building also cannot exceed the allowable height and area requirements outlined in Chapter 5 of the IBC. There is no exception to this limitation when following the prescriptive compliance path; however, some alternate methods are allowed under the work area method, which will be discussed later in this text.

Section 502.1 of the IEBC also states that the addition cannot make the existing building any less conforming to the requirements of the IBC than it was prior to the addition. As an example, when constructing an addition, the building owner will likely desire to have a new opening in the existing building to allow communication between the existing space and the new space. These alterations to the existing building will likely need to be addressed separately in accordance with Section 503 of the IEBC, while the addition itself would need to comply with Section 502 of the IEBC.

While the additions themselves must comply with the IBC, there are numerous triggers that would require items to be addressed within the existing building as a result of the addition. Those triggers and the associated rules for compliance are outlined below.

5.2.1 Disproportionate earthquake damage. This was previously discussed in Sec. 4.5.2 of this book in relation to repairs. If the structure is located within Seismic Design Category D, E, or F and portions of the lateral force-resisting systems (LFRS) have been damaged such that they have been reduced in capacity by as much as 10 percent, they should be considered as having "substantial structural damage." This requirement only occurs when the spectral accelerations at 0.3 s are less than 40 percent of the values at 0.2 s. To better understand disproportionate earthquake damage, refer to Sec. 4.5.2 of this book.

5.2.2 Flood hazard areas. Throughout the IEBC there is a trigger for existing facilities to be upgraded to meet the current code's requirements for construction in flood hazard areas. This was discussed in detail in Sec. 4.2 of this book, but in general, it occurs anytime the construction cost exceeds 50 percent of the market value. Section 502.3 of the IEBC requires the existing structure to meet the current flood design requirements of the IBC or *International Residential Code®* (IRC®) if the additions are considered "substantial improvements." If the addition is not considered a substantial improvement, then only the addition itself is required to meet the current IBC and IRC requirements in relation to flood design. Figure 4.7 of this book displays the difference between an addition triggering the substantial improvement requirements and an addition that does not.

5.2.3 Gravity structural elements. Section 502.4 of the IEBC limits the amount of new gravity loads that existing structural elements can resist without requiring an evaluation. Similarly, this section limits the amount an existing structural element can be reduced in capacity before an evaluation is required. To understand this rule, it is first important to understand the gravity loads that the design professional should address.

Chapter 16 of the IBC lists the gravity loads that need to be considered in design. These typically include dead loads, live loads, and snow loads. Dead loads are the actual weight of the materials in the structure, while Section 202 of the IBC defines live loads as loads produced by the use and occupancy of the structure. This can include the load of the persons, furniture, vehicles, and more. Snow loads are self-explanatory, and the IBC requires that they be determined in accordance with Chapter 7 of ASCE 7-16.

If the addition causes an increase in gravity loads to existing structural members of 5 percent or more, the existing structural element must be evaluated and made to comply with the current IBC requirements for new structural elements. Similarly, if the existing structural element has been decreased in capacity by 5 percent or more, an evaluation per the IBC must also be performed. This is known as the "5 percent rule."

The following are some examples of when the 5 percent rule might be triggered when dealing with an addition to an existing building:

- A vertical addition (as seen in Fig. 5.1) will increase the dead and live loads resisted by existing structural elements.
- An attached horizontal addition would likely produce additional gravity loads at the point of connection with the existing building. Examples would be attached decks, carports, patio covers, or even building additions.
- An attached horizontal addition where holes are cut through existing bearing walls to extend utilities or to create an opening between the existing and new.
- If an attic space is converted into living space, the conversion would likely require additional materials, which would increase the dead loads, and the live loads also would be increased significantly.
- Additional snow drift loads as shown in Fig. 3.6.

Figure 5.1 Vertical addition in Tallinn, Estonia. (*Wikimedia, 2008.*)

If the increased gravity loads do not exceed 5 percent of what the structural member was previously supporting, or the structural member has not been decreased in capacity by more than 5 percent, no action is required. The only exception to complying with the 5 percent rule is for Group R occupancies that have fewer than five dwelling or sleeping units and the building conforms to the conventional construction requirements of the IBC or IRC.

5.2.4 Lateral structural elements. Similar to the 5 percent gravity rule, there is a 10 percent lateral rule. Lateral structural elements are those elements that resist the wind and seismic loads imposed on the structure. These include diaphragms (e.g., roof and floor sheathing and fastening) and vertical members or systems (e.g., shear walls, moment frames, braced frames, cantilevered columns, and more). If any of the LFRS are required to support an increased load of 10 percent, or if they have been reduced in capacity by 10 percent or more, they must be evaluated to show full compliance with the wind and seismic provisions of the IBC. This is known as the "10 percent rule."

Figures 5.2*a* and 5.2*b* provide an example of when an addition might increase the loads to existing LFRS by more than 10 percent. In this example, the home owner is constructing a simple addition on the back of the home to create a larger kitchen space and a covered patio. The highlighted area in Fig. 5.2*a* shows the portion of the roof diaphragm that is supported by the northmost exterior wall, which also serves as a shear wall. Figure 5.2*b* shows that the same

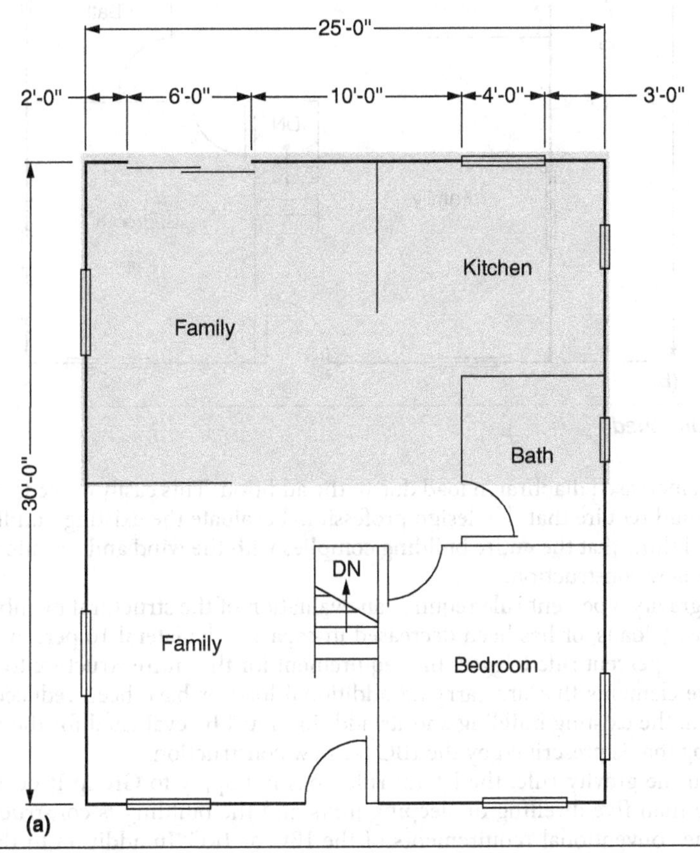

Figure 5.2 ***(a)* Roof diaphragm supported by northmost wall and *(b)* addition causing increased diaphragm loads.**

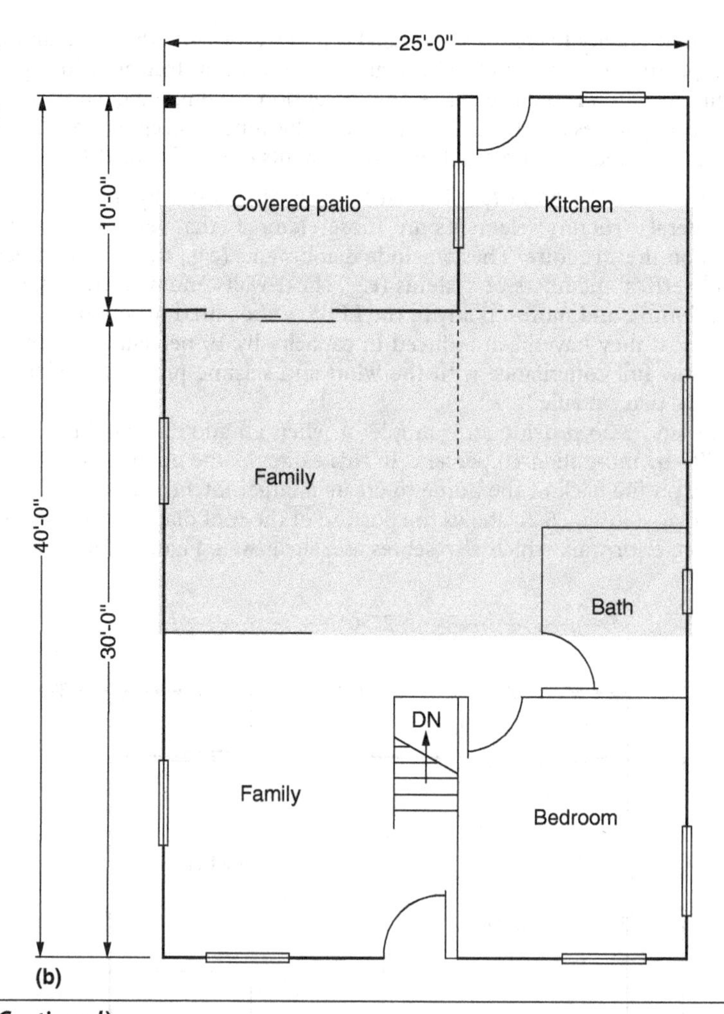

Figure 5.2 ***(Continued)***

wall carries an increased diaphragm load due to the addition. This easily exceeds the 10 percent trigger and would require that the design professional evaluate the existing building and addition as one and show that the entire building complies with the wind and seismic requirements of the IBC for new construction.

While the gravity 5 percent rule requires an evaluation of the structural member that is carrying new gravity loads, or has been decreased in capacity, the lateral 10 percent rule is much different. The 10 percent rule triggers the requirement for the entire structure to be evaluated, not simply the elements that are carrying additional load or have been reduced in capacity. Once triggered, the existing building and its addition must be evaluated for the full wind and seismic loading that is prescribed by the IBC for new construction.

Just as with the gravity rule, the lateral rule does not apply to Group R occupancies that have no more than five dwelling or sleeping units and the building is constructed in accordance with the conventional requirements of the IBC or IRC. In addition to this exception, the 10 percent lateral rule does not apply to additions that are built structurally independent of the existing building.

Figure 5.3 highlights two key elements that should be addressed if the owner will be constructing the addition independent of the existing building. The first item is to ensure that a sufficient gap is created between the existing building and the addition. The minimum gap must be calculated by the structural engineer of record (SER) in accordance with Section 12.12.3 of ASCE 7-16. The second item can be seen when looking at the footing in Fig. 5.3. Most footings are constructed concentrically, meaning the loads are applied at the center of the footing. When constructing near an existing foundation, it may not be possible to do this and would therefore require an eccentric footing as shown in this image. An eccentrically loaded footing requires additional analysis by the SER.

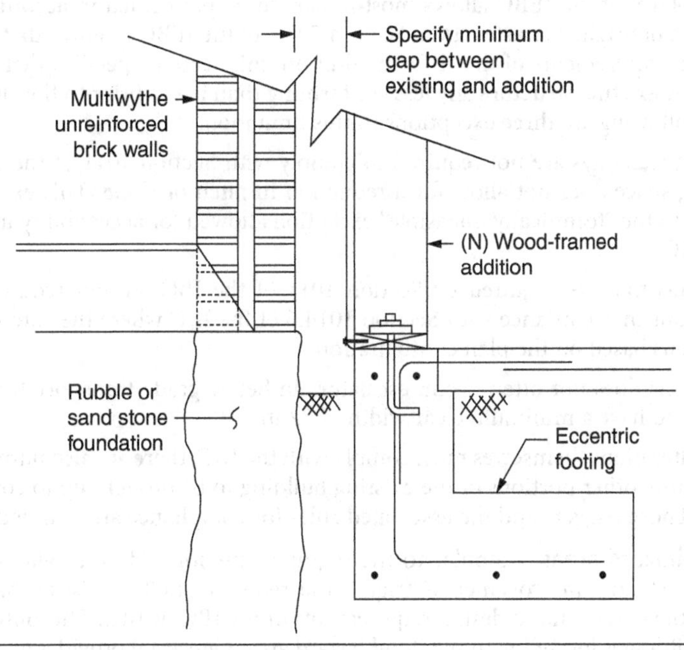

Figure 5.3 Structurally independent addition.

5.2.5 Smoke alarms. Section 502.6 of the IEBC requires that additions to all Group R and Group I-1 occupancies trigger the requirement for smoke alarms to be provided in accordance with Section 1103.8 of the *International Fire Code®* (IFC®). This trigger is not for the addition only, but for every portion of the existing building, including addition, that has Group R or Group I-1 occupancies. Section 1103.8 of the IFC requires that the smoke alarms be located in accordance with Section 907.2.10 of the IFC, that the alarms be interconnected, and defines the required power source for the alarms. This trigger is repeated in multiple locations throughout the IEBC.

5.2.6 Carbon monoxide alarms. Section 502.7 of the IEBC requires that additions to all Group R, I-1, I-2, and I-4 occupancies trigger the requirement for carbon monoxide alarms to be provided. These shall be installed in existing dwelling and sleeping units as specified in Section 1103.9 of the IFC. This trigger is repeated in multiple locations throughout the IEBC.

5.2.7 Additions to Group E. If an addition is made to an existing Group E occupancy, and the occupant load of that addition is greater than 50, Section 502.8 of the IEBC requires that a storm shelter be constructed in accordance with ICC® 500. The storm shelter is only required when the existing building does not already have a storm shelter and the site is subject to design wind speeds associated with tornadoes of 250 mi/h or more.

5.3 *Alterations*

Section 202 of the IEBC defines alterations as "any construction or renovation to an existing structure other than repair or addition." Alterations encompass most of the work that is performed to existing buildings. Residential remodels or commercial tenant improvements would be lumped into this category. In addition, most repairs and additions often incorporate alterations in the work being performed. If that is the case, the owner must comply with the requirements for the repair or addition while also accounting for the requirements for the alterations.

While Chapter 4 of the IEBC allows most repairs to be performed in accordance with the code at the time of original construction, Section 503.1 of the IEBC requires that all alterations conform to the requirements of the IBC. In addition, this section specifies that the alteration cannot make the existing structure any less conforming than it was prior to the alteration. With that said, the following are three exceptions to this limitation:

- Existing stairways are not required to comply with Section 1011 of the IBC when the existing space does not allow for a reduction in pitch or slope. This exception is very similar to the "technically infeasible" exception allowed for accessibility in Chapter 3 of the IEBC.
- Handrails that are required by Section 1011 of the IBC are not required to provide extensions in accordance with Section 1014.6 of the IBC where the extension could be hazardous based on the plan configuration.
- While this does not often occur, escalators in below-grade transportation stations are allowed to have a minimum clear width of 32 in.

While the alterations themselves must comply with the IBC, there are also numerous triggers that would require other portions of the existing building to be brought up to comply with the current code. Those triggers and the associated rules for compliance are outlined below.

5.3.1 Flood hazard areas. Similar to the requirements for additions, alterations that are considered "substantial improvements" trigger the requirement for existing buildings to be brought up to the current flood design requirements of the IBC or IRC. This only occurs when the existing building is located within a flood hazard area. Figure 4.9 provides an example of an alteration that is a substantial improvement and therefore triggers this requirement.

5.3.2 Gravity structural elements. Just as with additions, Section 503.3 of the IEBC requires existing structural members that are supporting new loads or that have been decreased in capacity to be evaluated. This evaluation is only required when the added load, or decreased capacity, has triggered the 5 percent rule. Again, Group R occupancies that have fewer than five dwelling or sleeping units and that are constructed per the conventional construction requirements are exempt. In addition, Section 503.3 includes another exemption for adding a new roof covering that is no more than 3 lb/ft^2 so long as only one layer of roofing materials currently exists.

5.3.3 Lateral structural elements. There are three important items to remember in relation to alterations that are made to existing LFRS. The first item is that voluntary seismic improvements can be made so long as the requirements noted in Sec. 5.3.12 of this chapter are met. The second item is that the alteration cannot create a structural irregularity. If a structural irregularity is created, a complete analysis of the entire building will be required.

The third, and last, item is to remember the lateral 10 percent rule. If members of the LFRS are reduced in capacity or if the load is increased by more than 10 percent, then an evaluation is required. As noted in Sec. 5.2.4 of this chapter, the evaluation is not for the individual member that was affected, but for the entire structure. The key difference between the evaluation required by Section 503.4 of the IEBC and what is required for additions is that the seismic portion of the analysis can be performed to the **reduced seismic forces.** This allows the design professional to check the existing structure considering a 25 percent reduction in the seismic forces.

Figures 5.4*a* and 5.4*b* provide an example of when an alteration would reduce the capacity of an existing LFRS and therefore triggers the requirement for a complete evaluation of the building for full wind and **reduced seismic forces.** The reader will notice that the example provided is the same shown in Figs. 5.2*a* and 5.2*b*, which portrayed added loads due to an addition. As was noted previously, additions often require alterations to be made to the existing building as well. In this case, a new opening was created as part of the kitchen addition, which reduced the shear wall length by 53 percent. The owner of this building will need to provide the code official with a detailed evaluation of the building and will need to show that it meets 100 percent of the IBC-required wind and 75 percent of the IBC-required seismic forces.

5.3.4 Seismic Design Category F. If the alterations being performed encompass more than 50 percent of the building area and the project is located within Seismic Design Category F, the entire building must be evaluated to conform to the full wind and **reduced seismic** requirements of the IBC. This trigger will not be met too often, as very few buildings are assigned to Seismic Design Category F. Section 1613.2.5 of the IBC defines Seismic Design Category F as

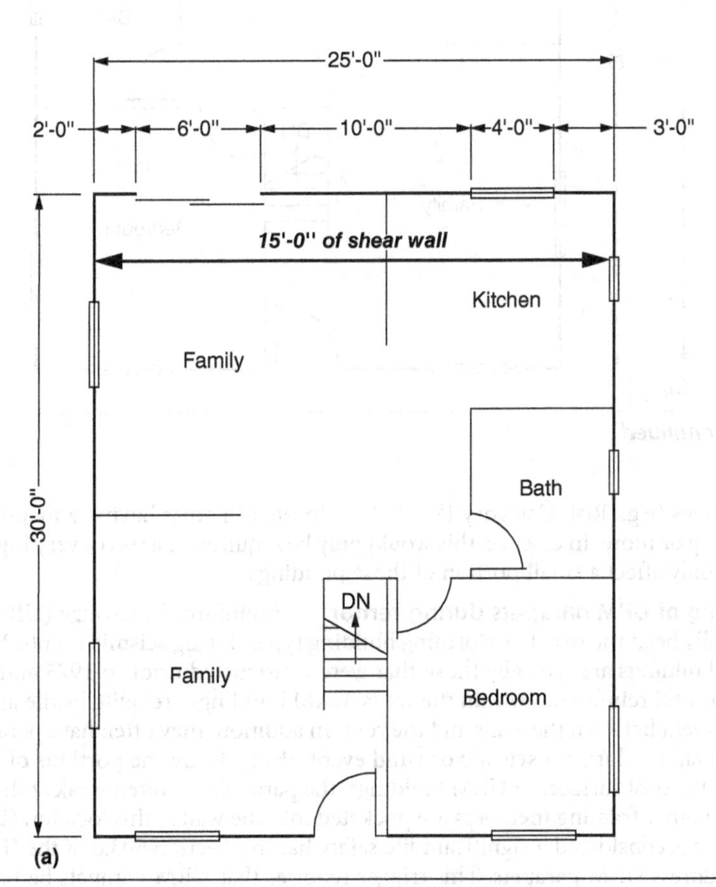

Figure 5.4 *(a)* **Existing shear wall length and** *(b)* **altered shear wall length.**

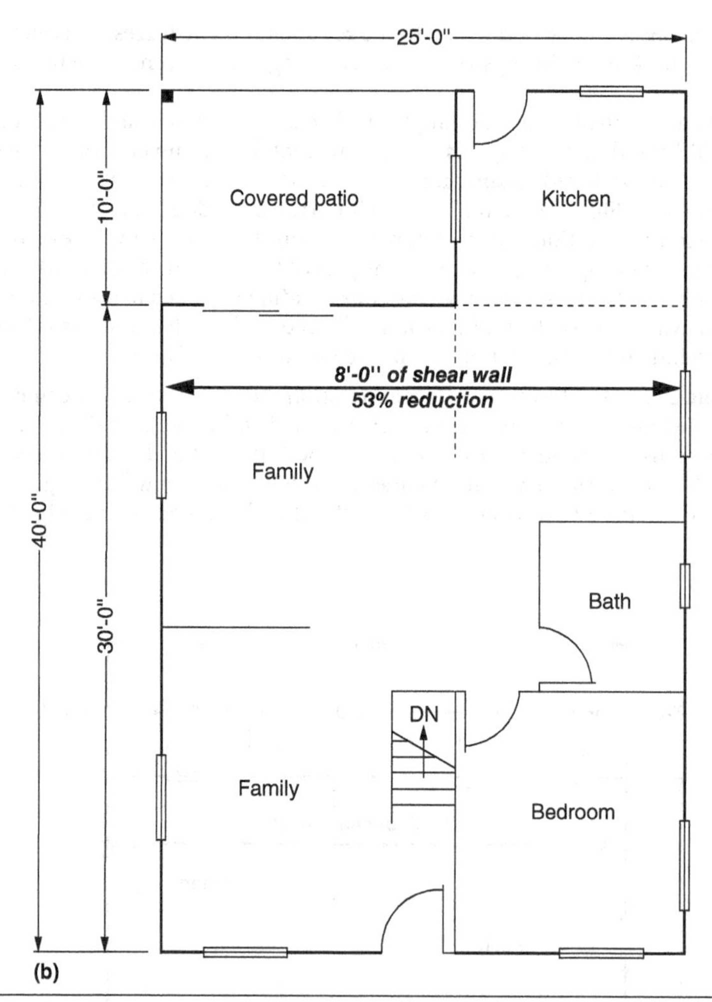

Figure 5.4 ***(Continued)***

essential facilities (e.g., Risk Category IV) that are located in areas having a mapped 1-s acceleration of 0.75 g or more. In essence, this would only be required in areas of very high seismicity, and it would only affect a small portion of those buildings.

5.3.5 Bracing of URM parapets during reroof. Unreinforced masonry (URM) buildings have historically been the worst-performing building types during seismic events. In the United States, URM buildings are typically those that were constructed prior to 1975 and they do not have adequate steel reinforcing within the walls. URM buildings are quite brittle and often lack proper attachment between the walls and the roof. In addition, they often have parapets that are susceptible to failure during a seismic or wind event. Parapets are the portions of the wall that project above the roof surface. In URM buildings, the parapets are often weak at the roofline, as roof joists and other framing members are pocketed into the wall at this location (USSC, 2016).

Because this is considered a significant life safety hazard, Section 503.6 of the IEBC includes a trigger to address URM parapets. This trigger requires that URM parapets be braced against lateral loads (e.g., wind and seismic) when at least 25 percent of the roof area is being reroofed.

This does not require a complete tear-off of the roof but could be as minor as adding a new layer of shingles to just a portion of the roof. The evaluation can be performed using the **reduced seismic loads**, and the bracing is required to resist the out-of-plane seismic forces that are calculated.

Figure 5.5*a* shows what URM parapet bracing might look like installed, while Fig. 5.5*b* displays a common detail that might be included in the construction plans for the work (USSC, 2016). There are many jurisdictions throughout the United States that require very little in terms of documentation when a permit application is submitted for a reroof.

As the actual reroof is the trigger for this requirement, it is recommended that jurisdictions in areas of high seismicity require additional documentation prior to issuing a reroof permit. As an example, the Structural Engineers Association of Utah (SEAU) created a "Reroofing Questionnaire" that was presented to code officials throughout the State of Utah (SEAU.org, 2018). If required by the local jurisdiction, the questionnaire requires the permit applicant to note if the building was constructed prior to 1975 and if URM parapets exist, in addition to a number of other items. If any of the items are answered in the affirmative, it flags the project, and the code official can require construction documents showing how the parapet bracing will be performed during the reroofing operations. All jurisdictions that have URM buildings and are located within high-seismic regions are encouraged to do something similar to this prior to issuing reroof permits.

(a)

Figure 5.5 **(*a*) URM parapet bracing—real-world application. (*Courtesy of FEMA.*) (*b*) URM parapet bracing—construction detail. (*USSC, 2016.*)**

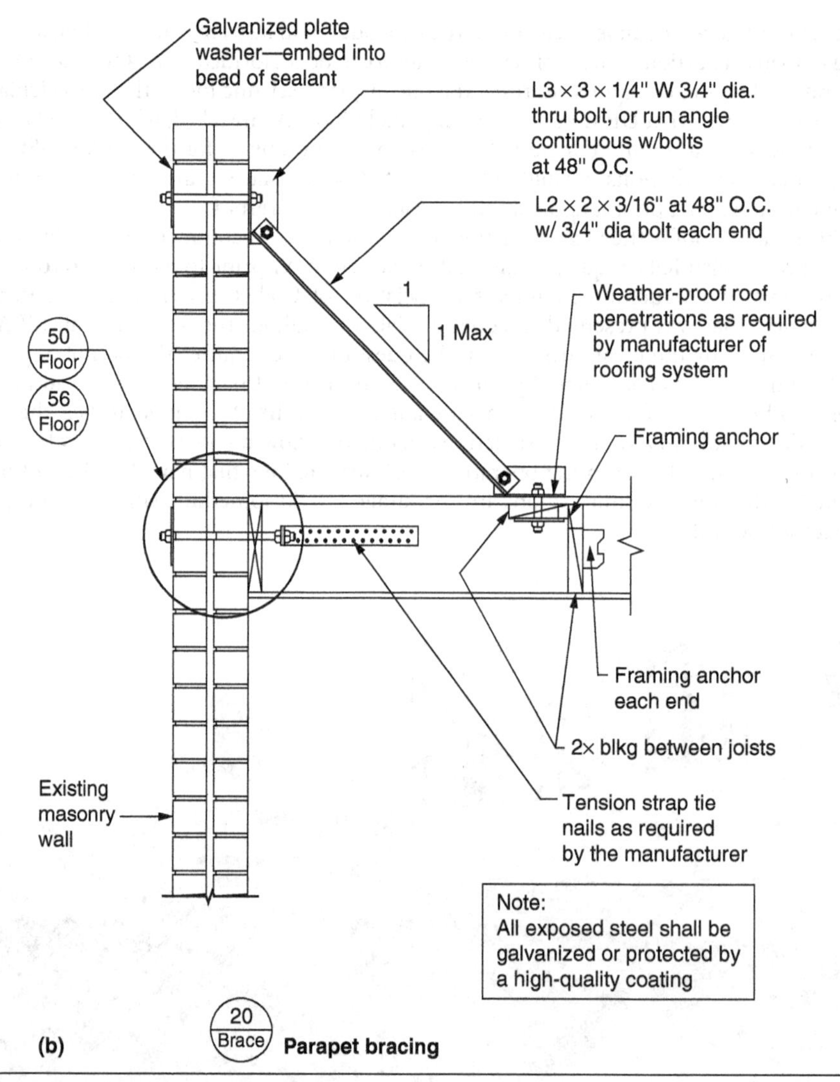

Figure 5.5 *(Continued)*

5.3.6 Anchorage of concrete or masonry walls. While URM buildings perform poorly during earthquake events, so do many concrete and reinforced masonry buildings. Those that are constructed in accordance with current codes should perform well, as they have a sufficient amount of reinforcing steel and detailing to provide the ductility needed, in addition to providing appropriate connections to maintain the lateral load path. These items are often lacking in older concrete and masonry buildings.

A good example of older concrete buildings that should be of concern are early tilt-up buildings. During the San Fernando earthquake in 1971 many of the early tilt-ups were found to have inadequate roof-to-wall connections. This caused many of the roofs to separate from the walls and led to collapse of several tilt-up buildings. Similarly, in the 1994 Northridge earthquake there were approximately 400 tilt-up buildings that were severely damaged. Because of these events, the 1997 *Uniform Building Code* (UBC) required an increase in wall anchor design forces and additional detailing requirements (McCormick, 2003).

The main concern with early concrete and reinforced masonry buildings is holding the building together so that occupants can escape the building. Because of this, Section 503.7 of the IEBC requires that an evaluation be performed of the roof-to-wall connection. This is only required if the project is located in Seismic Design Category C, D, E, or F and only when the alterations being performed affect 50 percent or more of the building area. It is also important to note that this is only required when flexible roof diaphragms are provided, as it is the connection of the roof diaphragm and the concrete or masonry wall that is of concern. The evaluation can be performed in accordance with the **reduced seismic forces**, but if the evaluation shows noncompliance, new roof-to-wall anchors must be installed.

5.3.7 Anchorage of URM walls. It is now important to combine the requirements of Secs. 5.3.5 and 5.3.6. These sections have highlighted that URM buildings are not ideal during seismic events and that the roof-to-wall connection is key to limiting the potential for collapse and allowing the occupants to get out of the building. With that said, like early concrete and reinforced masonry buildings, Section 503.8 of the IEBC requires the roof-to-wall connection in URM buildings to be evaluated.

This also is only for areas in Seismic Design Category C, D, E, or F and only when the alteration work affects more than 50 percent of the building area. Similarly, the evaluation can be performed to the **reduced seismic forces,** and new roof-to-wall anchors are only required if the evaluation shows noncompliance. Figure 5.6 provides an example of a new roof-to-wall connection added to URM residences. This is one of many such details that are provided in "The Utah Guide for the Seismic Improvement of Unreinforced Masonry Dwellings" (USSC, 2016).

5.3.8 Bracing of URM parapets due to alteration. While Section 503.6 of the IEBC requires URM parapets to be braced when reroofing, Section 503.9 of the IEBC also requires URM parapets to be braced when there is a significant alteration made to the building. While the reroofing trigger occurs when more than 25 percent of the roof area is affected, the alteration trigger is when more than 50 percent of the building area is being affected as part of the alteration. Section 503.9 also requires that parapet bracing be added in Seismic Design Category C, while the reroof trigger starts with projects in Seismic Design Category D. Once 50 percent or more of the building area is affected, an evaluation must be provided for the parapet bracing and the IEBC allows this to be performed using **reduced seismic forces.**

5.3.9 Anchorage of URM partitions. The 2018 IEBC has added a new requirement for URM partition walls. Partitions are considered nonbearing or nonstructural walls and occur throughout most buildings. Older concrete frame buildings often used URM to create infill walls between columns. These URM infills are intended to be nonstructural; however, during an earthquake they affect the overall response of the building. FEMA E-74 highlights the need to either isolate or remove heavy partitions, and especially URM infill walls, so that they do not adversely affect the seismic response of the building (FEMA E-74, December 2012).

Section 503.10 of the IEBC requires URM partitions to be removed or altered to resist out-of-plane seismic forces when more than 50 percent of the building area is to be altered. This requirement is only triggered in Seismic Design Category C, D, E, or F *and* when the partition in question is within the work area or is adjacent to an egress path. As many of these infill walls occur on the exterior of the building, it is important to consider the exterior egress path as well as the interior path.

5.3.10 Substantial structural alteration. Section 202 of the IEBC defines "substantial structural alterations" as those alterations that modify gravity load–carrying members that support at least 30 percent of the floor or roof area within a 5-year period. That is a pretty significant alteration, and the 2018 IEBC has a new requirement for buildings that undergo such an alteration. Section 503.11 of the IEBC requires buildings that undergo a "substantial

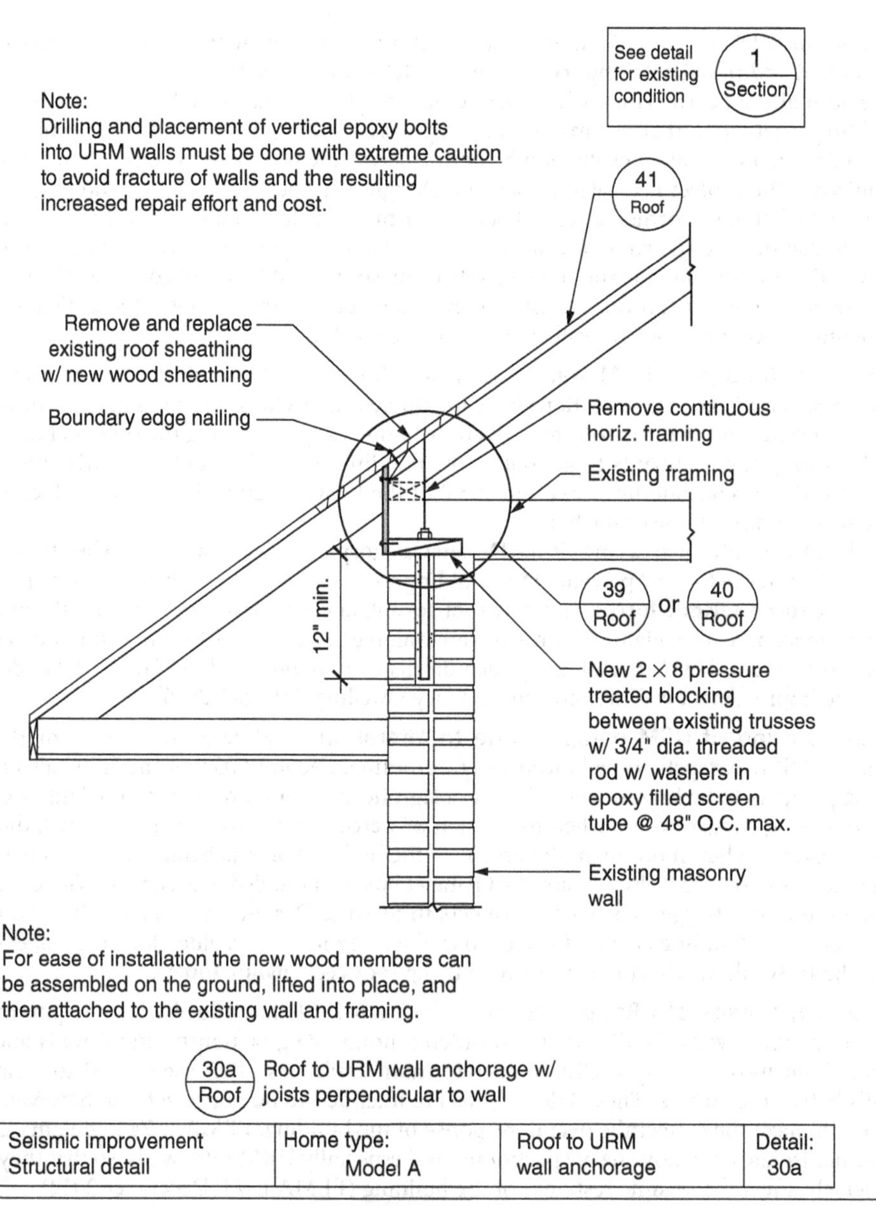

Seismic improvement Structural detail	Home type: Model A	Roof to URM wall anchorage	Detail: 30a

Figure 5.6 URM roof-to-wall connection. ***(USSC, 2016.)***

structural alteration," and where the altered space affects more than 50 percent of the building area, undergo an evaluation to show compliance to full wind and **reduced seismic forces.**

There are two exceptions to this requirement. The first is for Group R occupancies that have five or fewer sleeping or dwelling units and meet the conventional construction requirements of the IBC or IRC. The second does not require the entire building to be evaluated if the alteration solely occurs at the lowest level. In this case, only the LFRS at that level needs to be evaluated. If an evaluation is triggered by Section 503.11, all deficiencies identified must be remedied.

5.3.11 Roof diaphragms in high wind. Not all the lateral triggers in the IEBC are for seismic purposes. Section 503.12 of the IEBC requires roof diaphragms to be evaluated when the ultimate design wind speed is greater than 115 mi/h. The trigger for this evaluation is not only the design wind speed, but also requires that at least 50 percent of the roofing materials have been removed. In essence, this trigger would only come about when a roof tear-off is occurring and not when new roofing materials are being placed over existing materials.

The evaluation required by this trigger must consider the roof diaphragm itself, connections of the diaphragm to roof framing members, and the roof-to-wall connections. This evaluation is allowed to consider just 75 percent of the IBC-mandated wind loads, including uplift. If the current diaphragm and its connections are not capable of resisting 75 percent of the IBC-required wind loads, they shall be either replaced or strengthened. Figure 5.7 displays a detail that might be provided for strengthening the diaphragm itself and the connections to the roof framing members (USSC, 2016).

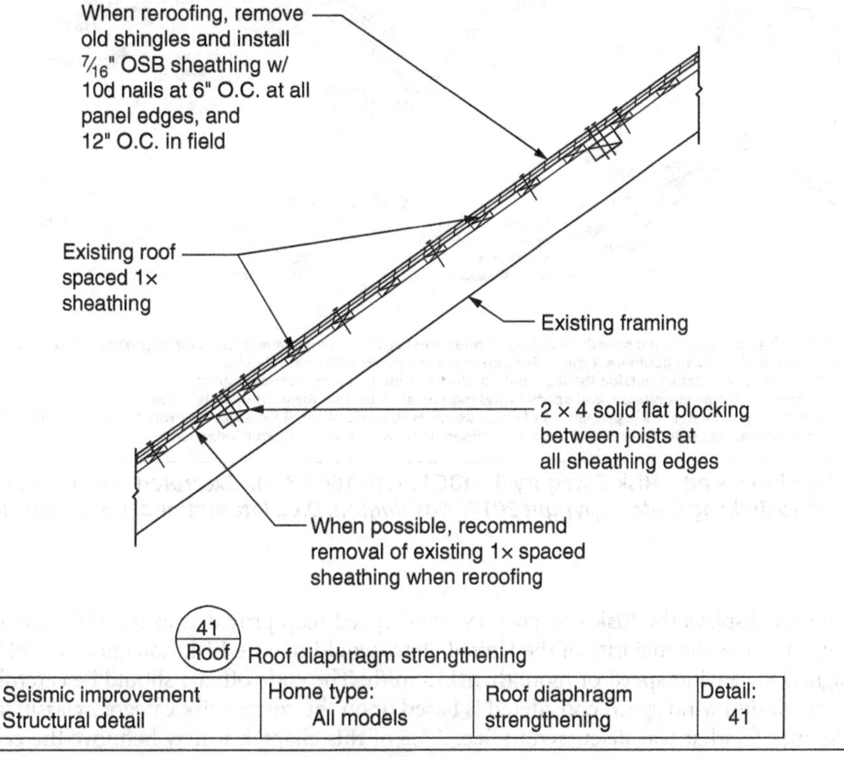

Figure 5.7 **Roof diaphragm strengthening. (*USSC, 2016.*)**

At this time, it is important to address the 115 mi/h wind design speed trigger. Figure 5.8 displays the design wind speeds for the continental United States for Risk Category II structures as provided in Figure 1609.3(1) of the IBC. Figure 5.8 shows that the majority of the United States, except for areas along the East Coast, would have a design wind speed of less than 115 mi/h.

Most code officials may feel that the 115 mi/h trigger has not been met by simply reviewing the map in Fig. 5.8; however, that would be a mistake. While most buildings would be classified as Risk Category II, Risk Category III and IV structures must be designed to a higher wind design speed.

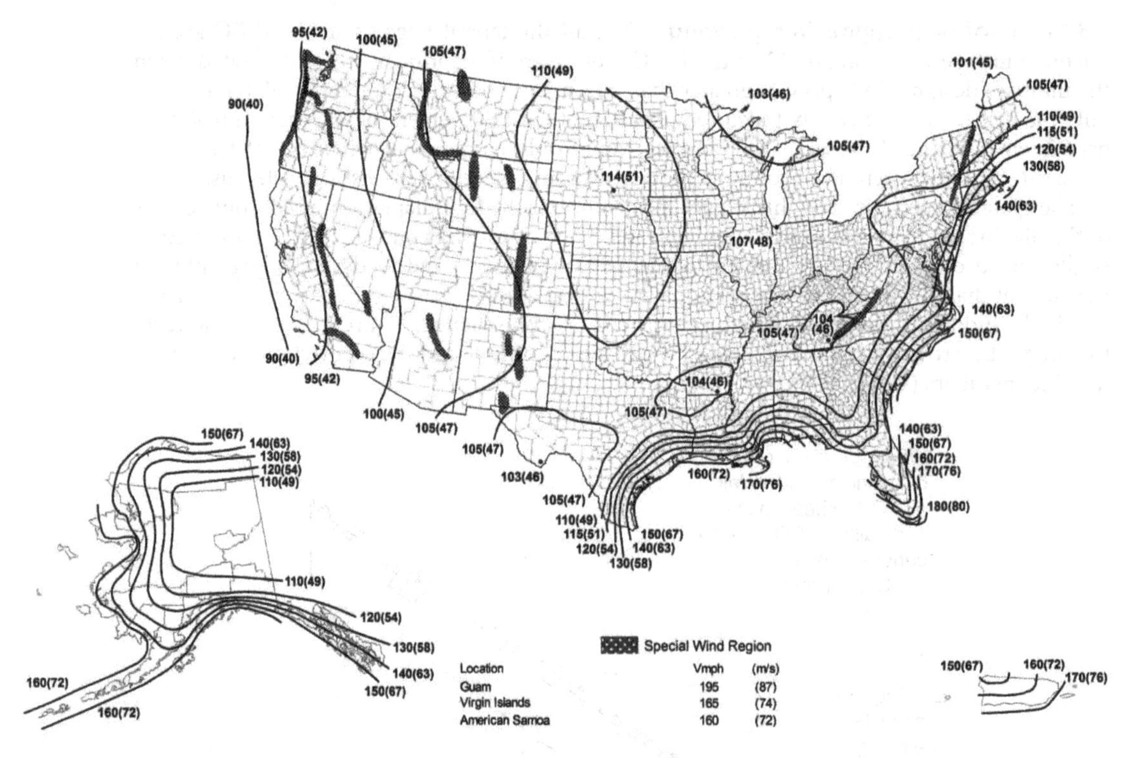

Notes:

1. Values are nominal design 3-second gust wind speeds in miles per hour (m/s) at 33 ft (10m) above ground for Exposure C category.
2. Linear interpolation is permitted between contours. Point values are provided to aid with interpolation.
3. Islands, coastal areas, and land boundaries outside the last contour shall use the last wind speed contour.
4. Mountainous terrain, gorges, ocean promontories, and special wind regions shall be examined for unusual wind conditions.
5. Wind speeds correspond to approximately a 7% probability of exceedance in 50 years (Annual Exceedance Probability = 0.00143, MRI = 700 Years).
6. Location-specific basic wind speeds shall be permitted to be determined using www.atcouncil.org/windspeed

Figure 5.8 **Wind design speed—Risk Category II—IBC Figure 1609.3(1).** *(Excerpted from the 2018 International Existing Building Code; copyright 2017. Washington, D.C.: International Code Council.)*

Figure 5.9 displays the Risk Category IV wind speed map provided in the IBC. The reader will notice that now the majority of the United States would require Risk Category IV buildings to be designed for a wind speed of more than 115 mi/h. The code official should be careful to ensure that the design wind speed considered is based upon the correct risk category classification.

Similar to what was discussed in Sec. 5.3.5 of this chapter, it may behoove the code official to develop a reroofing questionnaire that could catch these wind triggers prior to issuing a reroofing permit. The questionnaire could clarify the risk category, and therefore the required wind design speed, as well as the amount of tear-off that is occurring. This could raise red flags for the code official, and they would know when to require an evaluation of the roof diaphragm and its connections.

5.3.12 Voluntary LFRS upgrade. Up until this point the text has discussed when a trigger has been met that mandates an evaluation or upgrade to the existing building. At times, the building owner will choose to perform upgrades voluntarily, without a specific code mandate. Section 503.13 of the IEBC allows the building owner to perform a voluntary seismic improvement to a building so long as the work being performed does not make the building any less conforming.

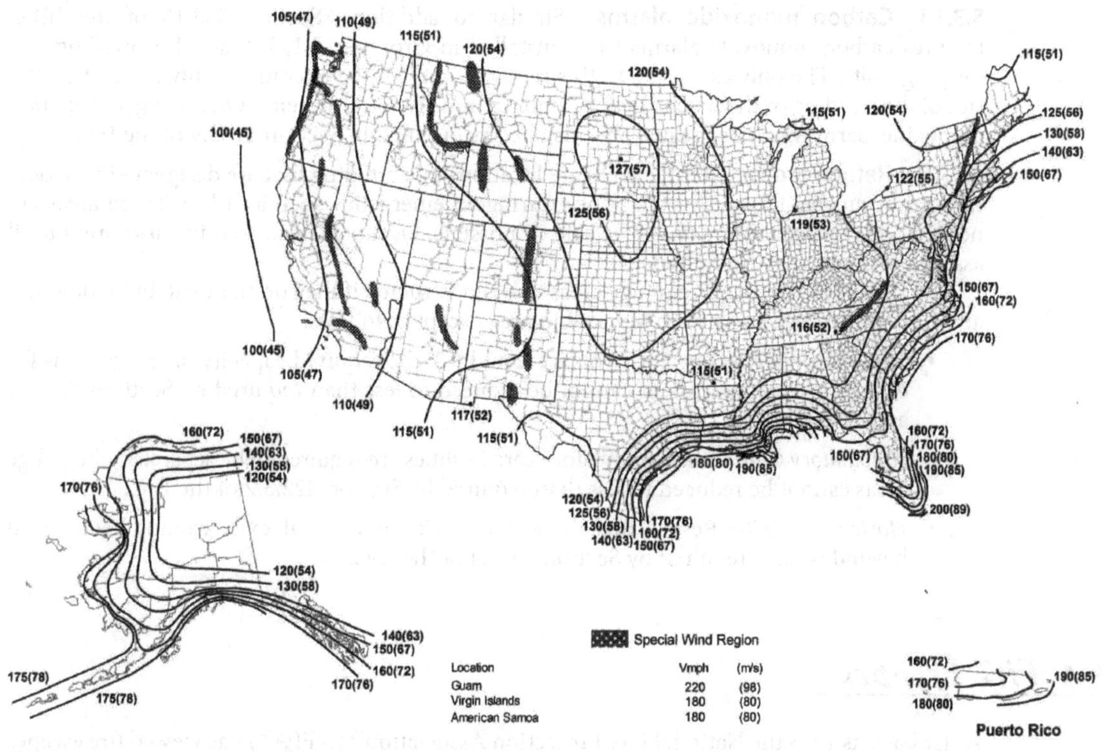

Notes:
1. Values are nominal design 3-second gust wind speeds in miles per hour (m/s) at 33 ft (10m) above ground for Exposure C category.
2. Linear interpolation is permitted between contours. Point values are provided to aid with interpolation.
3. Islands, coastal areas, and land boundaries outside the last contour shall use the last wind speed contour.
4. Mountainous terrain, gorges, ocean promontories, and special wind regions shall be examined for unusual wind conditions.
5. Wind speeds correspond to approximately a 1.6% probability of exceedance in 50 years (Annual Exceedance Probability = 0.00033, MRI = 3000 Years).
6. Location-specific basic wind speeds shall be permitted to be determined using www.atcouncil.org/windspeed

Figure 5.9 Wind design speed—Risk Category IV—IBC Figure 1609.3(3). ***(Excerpted from the 2018 International Existing Building Code; copyright 2017. Washington, D.C.: International Code Council.)***

If the building owner chooses to perform a voluntary seismic upgrade, they are still required to provide an analysis to the code official; however, this analysis is not required to meet the **full seismic** or **reduced seismic** criteria discussed throughout this book. The analysis provided must show the following:

- That the capacity of the existing structural elements will not be reduced.
- That new structural elements are designed, detailed, and connected to the structure as required by the IBC for new construction.
- If nonstructural elements are removed and replaced, they shall be appropriately restrained in accordance with the IBC and Chapter 13 of ASCE 7-16.
- The alterations made do not create a structural irregularity as defined in ASCE 7-16.

5.3.13 Smoke alarms. Just as with additions, Section 503.14 of the IEBC requires smoke alarms to be provided when alterations are made to individual sleeping and dwelling units in Group R or I-1 occupancies. These alarms are to be located and installed as outlined in Section 1103.8 of the IFC.

5.3.14 Carbon monoxide alarms. Similar to additions, Section 503.15 of the IEBC requires carbon monoxide alarms to be installed in Group R, I-1, I-2, and I-4 dwelling and sleeping units. The only exception to this requirement is if the alterations only affect the exterior of the building or the mechanical, plumbing, and electrical systems. Once triggered, carbon monoxide alarms must be located and installed as outlined in Section 1103.9 of the IFC.

5.3.15 Refuge areas. Refuge areas are locations in a building that are designated as a location for occupants to wait for assistance during an emergency such as a fire. These areas are needed when evacuation may not be safe or possible. Occupants can wait in these areas until assisted or rescued by firefighters.

The IEBC does not require refuge areas to be constructed if they do not exist, but it does not allow existing refuge areas to be reduced in area except as follows:

- *Smoke Compartments:* For Groups I-2 and I-3, the required capacity of refuge areas for each smoke compartment cannot be reduced to less than required by Sections 407.5.1 and 408.6.2 of the IBC.
- *Ambulatory Care:* When ambulatory care facilities are required to be separated, the refuge areas cannot be reduced to less than required by Section 422.3.2 of the IBC.
- *Horizontal Exits:* Rescue areas associated with horizontal exits cannot be reduced beyond what is required by Section 1026.4 of the IBC.

5.4 *Fire Escapes*

As far back as 1913 the National Fire Protection Association® (NFPA®) has viewed fire escapes as a problematic solution to getting people safely out of buildings. NFPA's Committee on Safety to Life found that fire escapes have a number of common defects, including inaccessibility, tendency to be unshielded against fire, poor design, absence of ladders or stairs to ground level, ice and snow coverage, and much more. The 2012 fire code included significant updates to clarify the need to provide inspection, testing, and maintenance of existing fire escapes (Baldassarra, 2014.).

Due to their numerous hazards in relation to egress, the IBC does not allow fire escapes to be constructed as part of the means of egress for new facilities. With that said, Section 504.1.3 of the IEBC allows for new fire escapes to be constructed at times. This only occurs when a second means of egress is required from an existing space, and exterior stairs cannot be constructed due to site limitations. New fire escapes cannot incorporate ladders, nor can they access the escape through a window.

At no time can the fire escapes serve as more than 50 percent of the required exit capacity. All new fire escapes, as well as all existing fire escapes, must comply with the design provisions outlined in Section 504 of the IEBC. This defines where escapes can be located, the construction limitations, dimensional requirements, and prescribes opening protective requirements as described below.

5.4.1 Location. When the fire escape is in the front of the building, the lowest landing cannot be closer than 7 ft from grade nor higher than 12 ft from grade. In addition, it must be equipped with a counter-balanced stairway to the street. If located within an alleyway, or next to streets that are less than 30 ft wide, the clearance to the lowest landing must be at least 12 ft.

5.4.2 Construction. Similar to stairways, fire escapes shall be designed to support a live load of 100 lb/ft^2. In addition, they should be constructed of noncombustible materials. With that said, if attached to buildings of Type V construction, the escape can be constructed utilizing no less than 2 in. nominal lumber. In addition, if the building is of Type III or Type IV construction and the building includes a combustible roof, the guardrails and walkway surface can be constructed of no less than 2 in. nominal lumber.

5.4.3 Dimensions. Stairways cannot be less than 22 in. wide with tread depths and rise not less than 8 in. The landings themselves cannot be less than 40 in. wide by 36 in. long. In addition, the landing cannot be located more than 8 in. below the door.

5.4.4 Opening protectives. Unless the building is fully sprinklered, all doors and windows within 10 ft of the fire escape stairways must be protected with 0.75-h opening protectives as outlined in Chapter 7 of the IBC.

5.5 *Window Replacement*

Section 505 of the IEBC addresses what is required when replacing windows in existing buildings. In general, Section 505.1 states that replacement glass must meet the requirements for new installations. That includes adding protected glazing in hazardous locations or fire-resistance-rated glazing within fire-rated assemblies as is outlined in the IBC or IRC. In addition to complying with the requirements for new installations, the IEBC highlights two other requirements for replacement windows as described below.

5.5.1 Opening control devices. All replacement windows in buildings with Group R-2 or R-3 dwelling or sleeping units, or one- and two-family residences and townhomes regulated by the IRC, shall have opening control devices installed where all of the following apply:

- The window is operable and when fully opened allows for the passage of a 4-in. diameter sphere.
- The replacement includes the sash and frame.
- *Group R-2 or R-3:* The top of the sill is less than 36 in. from the floor.
- *Regulated by IRC:* The top of the sill is less than 24 in. from the floor.
- The vertical distance from the window sill to grade level is greater than 72 in.

If triggered, the opening control device cannot limit the clear opening area of the window any more than allowed by Section 1030.2 of the IBC. The IEBC provides only two exceptions that would not require the installation of an opening control device if triggered by the five bullets listed above. Both cases do not require the opening control device when the replacement windows are provided with fall protection devices. For windows located 75 ft above grade, the fall protection device must comply with ASTM F2006, while windows located closer to grade and incorporating fall protection devices are require to comply with ASTM F2090.

5.5.2 Emergency escape and rescue openings. Section 505.3 of the IEBC addresses the requirements for emergency escape and rescue openings. Like opening control devices, this section applies to Group R-2 and R-3 occupancies as well as one- and two-family residences and townhomes that are regulated by the IRC.

In recent years, the IEBC verbiage has been revised to clarify that existing windows within these types of occupancies do not need to be increased in size to meet the dimensional requirements for emergency escape and rescue openings required by the IBC or IRC. Rather, there are simply two conditions that must be met if a replacement window is installed in what would be an emergency escape and rescue opening per the IBC or IRC. Those two conditions are as follows:

- Replacement windows must be the largest standard size window that is made to fit the existing window opening, and

- The operable portion of the replacement window must provide a window opening area that is equal to or greater than the existing window.

The intent of this provision in the IEBC is to not discourage or prevent improvements in older residences by requiring replacement windows to fully conform to the current code (DWM, 2012). It should be clarified that this provision is for "alterations" only. Should the existing building be undergoing a change of occupancy, the replacement window would need to comply with the provisions of the IBC or IRC for emergency escape and rescue openings.

Section 505.4 was added to the 2018 IEBC and provides some additional provisions for emergency escape and rescue openings. This section requires that the openings be operational from the inside of the room and that no special tools or knowledge is required to operate these openings. It also notes that they should not require the use of a key or excessive force and that grates or bars shall not reduce the net clear opening provided. The last item that is included is that the smoke alarms are to be provided in accordance with Section 907.2.10, regardless of the scope of the alteration being made.

5.6 *Change of Occupancy*

The definition provided in Section 202 of the IEBC notes that a change of occupancy can occur within a building under three separate scenarios. The following provides a brief description of each scenario described in the definition along with an example for each.

5.6.1 Change of occupancy classification. Consider that an existing warehouse (Group S-2) is being converted to a conference center (Group A-3). It is changing from one classification (e.g., S) to another (e.g., A).

5.6.2 Change from one group to another. Consider a building supporting a glass blowing business (Group F-2) is being improved to support a facility that will manufacture boats (Group F-1). While both are within the F classification, the specific code requirements fall under different groups.

5.6.3 Change in use within a group that triggers change in application of the code. Consider a tenant space within an office building (Group B) that is undergoing a tenant improvement to support a new ambulatory care center. While an ambulatory care center falls under Group B, the code has several additional requirements for an ambulatory care occupancy that are not required for most Group B occupancies, such as increased electrical requirements under the *National Electrical Code®* (NEC®).

If any of the above-noted items are triggered, the building is undergoing a change of occupancy, and the requirements of Section 506 of the IEBC must be met for the existing building. In general, Section 506.1 of the IEBC requires all changes of occupancy to comply with the provisions of the IBC for the specific use or occupancy, with the exception of the structural provisions, which will be addressed in a moment. The only way out of complying with the IBC as part of the change of occupancy is if the code official is willing to acknowledge that the change of occupancy is less hazardous than the existing occupancy.

So when can the code official acknowledge that the change is "less hazardous"? The IEBC states that this decision must be based upon life and fire safety. Section 5.6.2 of this chapter describes a change of occupancy from Group F-2 to Group F-1. This is from a "low-hazard" occupancy to a "moderate-hazard" occupancy. This obviously does not meet the exception, but the code official might waive the requirement if it were reversed and a Group F-1 occupancy was changed to a Group F-2 occupancy.

Section 506.1.1 of the IEBC notes that if no change in the occupancy classification has occurred, only those items that the code official mandates are required to meet the IBC, rather than the entire structure. This would apply to the scenario described in Sec. 5.6.3, where the ambulatory care center will be added to the existing Group B space. In this instance, the code official can simply require the certain elements required for an ambulatory care facility, but which are not required for all Group B occupancies, to be updated to the current code.

In general, changes in occupancy are required to comply with the IBC, as has been noted above. To this point, the only exceptions discussed occur when the use is considered to be "less hazardous" or when the occupancy classification does not change, in which case full compliance is not mandated. In addition to these two items, Section 506 of the IEBC has lesser requirements for stairways and structural items than is required for new construction. The following outlines these provisions.

5.6.4 Stairways. Section 506.3 does not require existing stairways to meet the requirements of Section 1011 in the IBC, where the existing space does not allow for a reduction in the pitch or slope. It should be noted that if the facility can accommodate a change to the stairway without requiring significant structural modification, it should be accomplished.

5.6.5 Structural. Rather than require existing buildings that undergo a change of occupancy to comply with all of the structural provisions of the IBC, the IEBC simply requires a check of the structural members in existing buildings under certain circumstances. Those circumstances are in relation to the structural items described below.

Live loads. Structural members supporting live loads from a change of occupancy must be checked to verify that they are able to support the live loads listed in Table 1607.1 of the IBC. The exception to this requirement is the "5 percent rule" that has been discussed previously. To understand this a bit better, the reader should consider the following two examples.

- *Example 1:* An existing office space (Group B) is being converted into a cafeteria area (Group A-2). In accordance with Table 1607.1 of the IBC, the floor live load will be increased from 50 to 100 lb/ft^2. This increase is significant and easily triggers the 5 percent rule. As such, the existing floor framing and its supports, including foundations, must be checked to meet the current IBC live load requirements based upon the new occupancy.
- *Example 2:* An older hotel building (Group R-1) is being converted into an apartment complex (Group R-2). In accordance with Table 1607.1 of the IBC, the live load requirements of the hotel are the same as the requirements for an apartment complex. As such, the 5 percent rule has not been triggered and the existing gravity framing does not need to be checked. The only exception to this is if specific areas will be used differently as part of the change of occupancy and would require a higher live load to be considered.

If an increased live load occurs due to the change of occupancy, an evaluation complying with Section 303.1 of the IEBC must be provided. All elements that are shown to be non-compliant must be modified or replaced in a manner that would provide compliance with the current IBC.

Snow and wind loads. If the change of occupancy causes the existing building to be classified as a higher risk category, the entire building must be evaluated to meet the increased snow and wind load requirements of the IBC. IBC Table 1604.5 describes when a building should be classified as either a Risk Category I, II, III, or IV building. Table 1.5-2 of ASCE 7-16 lists the snow, ice, and seismic importance factors that should be considered for each risk category, while IBC Figures 1609.3(1) through 1609.3(8) provide the wind speed design values based upon risk category. When using the importance factors provided in ASCE 7-16 or the wind maps in the IBC, the snow and wind design values must be increased for higher risk categories.

Consider a medium-sized retail building (Group M) that is converted to a church (Group A-3). The occupant load count for the new use is roughly 400 persons. Because of the primary use and occupant load count for the space, the building will need to be reclassified from a Risk Category II to a Risk Category III structure per Table 1604.5 of the IBC. This will require that the existing building be evaluated for the full wind and snow load requirements of a Risk Category III structure per the current IBC. Any items that are found to be noncompliant must be remedied.

It should be highlighted that this requirement does not get triggered if the change of occupancy affects less than 10 percent of the aggregate area of the building. As an example, consider an ambulatory care facility that is added to a tenant space within a large office building. This ambulatory care center is less than 10 percent of the aggregate area of the building, but it does include emergency surgery facilities that would classify it as a Risk Category IV building per Table 1604.5 of the IBC. The main building is classified as Risk Category II. Because the change of occupancy occurs in less than 10 percent of the space, the building is not required to be checked for increased snow and wind loads.

Seismic loads. Similar to snow and wind, when the building is classified to a higher risk category, a seismic evaluation must be provided showing compliance with the **full seismic** provisions of the IBC. This evaluation must be performed in accordance with Section 303.3.1 of the IEBC. All items that are found to be noncompliant must be remedied.

There are three exceptions to requiring the **full seismic** evaluation. The first is identical to what was described for increased snow and wind. If the change of occupancy affects less than 10 percent of the aggregate area of the existing building, the evaluation is not required.

The second exception does not require a seismic evaluation when the change of occupancy does not result in the building being reclassified as an essential facility (e.g., Risk Category IV) and the building is located in Seismic Design Category A or B. Note that even in areas of lesser seismic risk, if the change of occupancy requires the building to be classified as Risk Category IV, the seismic evaluation is required by the IEBC.

The third exception is in relation to URM buildings in Seismic Design Category A or B. It states that if the change of occupancy results in the building being reclassified as Risk Category III that the evaluation does not need to meet the **full seismic** requirements, but rather Chapter A1 of Appendix A can be used (e.g., reduced seismic requirements).

Access to Risk Category IV. Section 506.4.4 is new to the 2018 IEBC. This section is only triggered when a change of occupancy causes the building to be reclassified as an essential facility (e.g., Risk Category IV). The section requires adjacent buildings that provide "operational access" to the essential facility to comply with the snow, wind, and seismic requirements of the IBC. In essence, not only is the essential facility required to be evaluated due to the change of occupancy but also adjacent facilities.

To better understand this requirement, we should review Section 1604.5.1 of the IBC. This section addresses how multiple occupancies are affected by the risk category of one occupancy. It first states that structures that include occupancies classified to different risk categories should be designed in accordance with the higher risk category. The only exception to that requirement is if the spaces are structurally separated. Now, if the spaces are structurally separated but they share life safety components, they still must be designed to the higher risk category.

Consider a hospital building (Risk Category IV) and an adjacent medical office building (Risk Category II). If one of the required means of egress components for the hospital passes through the medical office building, the office building itself would then need to be designed as a Risk Category IV facility, just as the hospital.

This new requirement in the IEBC is based upon this section in the IBC; however, it is only triggered for essential facilities (Risk Category IV). To better understand when this might be required, assume that an emergency operations center (EOC) will be added to an existing tenant space. The surrounding spaces, or buildings, are classified as Risk Category II yet if they share life safety components with the EOC, they would also need to be evaluated to the full IBC snow, wind, and seismic requirements. This is very similar to what is required under Section 1604.5.1 of the IBC for new construction.

At the completion of the change of occupancy work, the code official must issue a certificate of occupancy for all buildings, or portions of buildings, that undergo the change in use. This certificate must comply with Section 110 of the IEBC and shall not be issued until after the space has been shown to comply with the requirements of the IBC for that use in addition to the items outlined in Section 506 of the IEBC.

5.7 *Historic Buildings*

This entire chapter has focused on specific triggers that require updates to current code requirements, or a less-restrictive version of the current code, when existing buildings are undergoing additions, alterations, or changes in occupancy. When it comes to historic buildings these triggers are not mandatory. In addition, Section 507.1 of the IEBC states that repairs to the predamage condition are not mandatory.

Under the prescriptive method approach, triggers involving historic buildings only occur when there are specific life safety hazards, when they are in flood hazard areas, or when specific structural requirements are triggered. The following outlines the specific requirements for each one of these triggers in relation to historic buildings.

5.7.1 Life safety hazards. Section 507.2 of the IEBC allows the code official to require elements of historic buildings that are deemed to be a "distinct life safety hazard" to be remedied. The problem with this requirement is that it is very open-ended and often requires the code official to visit the site and perform an evaluation of the facility prior to issuing a building permit for any work so that life safety hazards can be identified.

Because of the open-ended nature of this requirement, it is recommended that local jurisdictions enforce the requirements in Chapter 12 of the IEBC when dealing with historic buildings rather than what is laid out in Section 507.1. While Chapter 12 of the IEBC is part of the work area compliance method, it prescribes a method that requires the owner or design professional to identify deficiencies rather than placing the burden solely on the code official. This is further discussed in Chap. 8 of this book.

5.7.2 Flood hazard areas. In general, historic buildings that are in flood hazard areas must comply with the requirements outlined throughout the IEBC. That is to say, if the work being performed is considered to be a substantial improvement, the building shall be made to comply with the flood design requirements of the current IBC or IRC. Of course, there is always an exception, and the IEBC does provide a way out of this requirement if the building is truly considered historic.

Section 507.3 states that if the building is listed on the National Historic Register, is in and contributes to an approved historic district, or is designated as historic under a state or local historic preservation program, it is not required to be brought into compliance with the flood design requirements of the current code. In order to rely on this exception, the owner or design professional should provide supporting documentation to the code official showing that the building does in fact meet one of the historic designations outlined in Section 507.3 of the IEBC.

5.7.3 Structural. By default, Section 507.4 of the IEBC requires historic buildings undergoing additions, alterations, or changes in occupancy to comply with all the structural provisions that have been outlined in this chapter. That includes not only the 5 percent gravity and 10 percent lateral rules but also other elements such as parapet bracing, roof-to-wall connections, and much more.

Of course, there are some exceptions, but these exceptions do not waive as many of the structural requirements as one might expect. The first exception allows the code official to accept existing floor live loads. The second states that if substantial structural damage has occurred, the repairs can be made to bring it back to its original condition rather than upgrading the building to comply with the current code requirements. While both of these exceptions provide a significant benefit to the owner, notice that many structural triggers have not been exempted. That includes evaluations to full wind, snow, and seismic loads if the building is reclassified to a higher risk category as part of a change of occupancy or evaluations required as part of alterations or additions.

Figure 5.10 Wichita-Sedgwick County Historical Museum.

As an example, consider the building shown in Fig. 5.10. This is an image of the Wichita-Sedgwick Historical Museum in Wichita, Kansas. The building was constructed in 1892 and was originally used as the City Hall. For the purposes of this example, assume that the old City Hall (Group B) is being converted to the current Historical Museum (Group A-3). Due to the size of the building Table 1604.5 of the IBC would reclassify the building as Risk Category III as part of the change of occupancy. While the increased live loads may not need to be checked as a result of the exception, the building would still need to be evaluated to the full snow, wind, and seismic loads that are required by the IBC for Risk Category III buildings.

CHAPTER 6

WORK AREA METHOD

6.1 *General*

The work area compliance method provides perhaps the greatest amount of benefits to the building owner and design professional. It comprises Chapters 6 to 12 of the *International Existing Building Code*® (IEBC®) and includes 32 of the 81 pages of the main code text. While the prescriptive approach is perhaps the easiest route to take, it is much more conservative. When dealing with larger buildings or more complicated projects, the owner and the design professional should seriously consider following the work area approach.

The work area method follows a "proportional approach" to code compliance. This means that as more work is being performed, more requirements will be triggered. Figure 6.1 highlights how this proportional approach is accomplished within the work area compliance method. Alterations are divided into three separate categories, Level 1, Level 2, and Level 3. Level 1 alterations denote that minimal work is being performed and therefore few triggers are met. Level 2 alterations are a little more involved and trigger additional items beyond that of a Level 1 alteration. Level 3 alterations are much more involved, while changes in occupancy trigger the greatest amount of code requirements to be met.

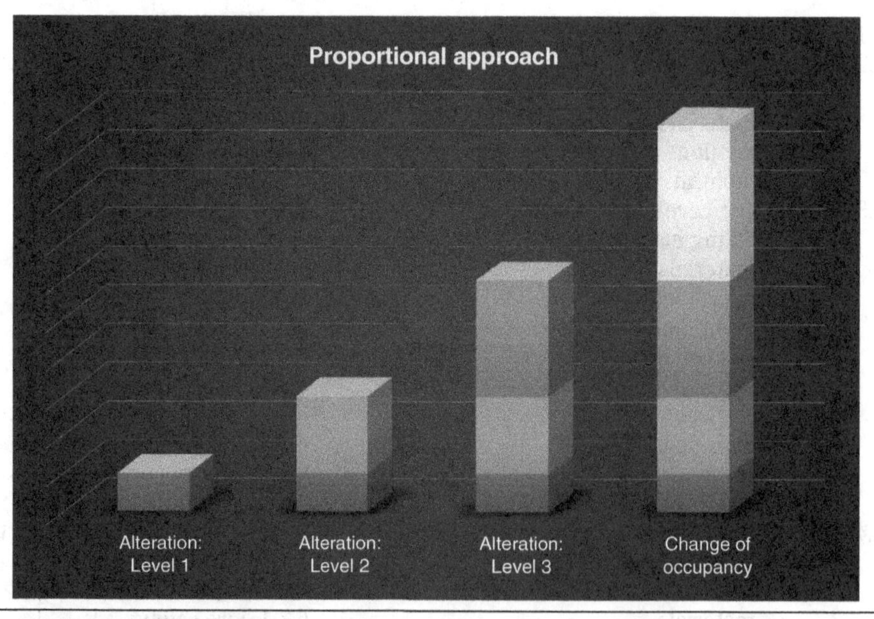

Figure 6.1 Proportional approach.

As shown in Fig. 6.1, the triggers met by Level 1 alterations are also required in Level 2 alterations. Each level builds upon the next. This is displayed as the color for Level 1 alterations is carried over for each subsequent level while additional items are added on the top. Changes in occupancy trigger the requirements of Level 1, Level 2, and Level 3 alterations while also including additional items to be met.

Section 601.2 requires that the construction documents identify the work area. This is repeated from Section 106.2.1 of the IEBC. There are multiple ways in which the design professional can clearly show the work area on the construction documents. It can be done by providing a layout of the building and simply hatching the area in which work is being performed, or oftentimes the designer will grayscale all existing conditions and use bold or dark lines to designate the locations where work is being performed.

As noted previously, the work area method builds upon itself and follows a proportional approach. The following sections will describe the base requirement and will slowly build upon that. The scope of each section will be identified, as well as the particular triggers that occur based upon the work being performed.

6.2 *Alterations: Level 1*

Chapter 7 of the IEBC prescribes the requirements for Level 1 alterations. Level 1 alterations include work of a minor nature. Section 602.1 of the IEBC describes these alterations as the removal, replacement, or covering of existing materials or elements. This could include reroofing operations, siding replacement, or the removal and replacement of rooftop mechanical equipment.

The alteration shall in no way make the building less safe than the existing condition. In fact, Section 701.2 states that if safety or sanitation is reduced as part of the alteration, it shall be shown to comply with the requirements of the current *International Building Code*® (IBC®). This would not be triggered very often on work that is being performed under a building permit but could often be the case for work being performed without a permit in place.

As has been stated many times throughout this text, if the Level 1 alteration is made to an existing building that is located within a flood hazard area and the work can be defined as a "substantial improvement," the entire building must be upgraded to meet the flood design requirements of the IBC or *International Residential Code*® (IRC®) for new construction. For Level 1 alterations it is highly unlikely that the "substantial improvement" trigger will be met. Simply adding new exterior cladding and reroofing, a building will not likely exceed 50 percent of the market value for the existing building. With that said, it is important that the reader remember this is a proportional approach, and the Level 1 requirements carry over to Level 2 and Level 3 alterations, as well as changes of occupancy. Should any of these alterations or changes of use exceed 50 percent of the market value of the existing building, it is required to meet the current code.

Like Section 505.4 of the prescriptive path, Section 701.4 was added to the 2018 IEBC for all existing buildings undergoing alterations. This section requires that emergency escape and rescue openings be operational from the inside of the room and that no special tools, knowledge, or keys are required to operate the openings. It also notes that the opening should not require excessive force, grates or bars shall not reduce the net clear opening, and smoke alarms are to be provided in accordance with Section 907.2.10 of the IBC.

Sections 703 and 704 of the IEBC state that the alterations made cannot cause a reduction in the existing fire protection or means of egress systems. In addition, Section 707 of the IEBC states that the energy conservation requirements of the current *International Energy Conservation Code*® (IECC®) or IRC are only required for the alteration itself and that alterations do not cause the entire building or structure to be upgraded to the current energy regulations.

Besides the items noted above, the main triggers associated with Level 1 alterations are in relation to building materials, reroofing, and structural requirements. The following subsections address what is specifically required by Chapter 7 of the IEBC in relation to each of these items. The reader will notice that in most cases the provisions within the work area method are duplicates of what has been specified within the prescriptive compliance approach. The main difference will be the trigger. Whereas the prescriptive approach will cause even small alterations to trigger significant code requirements, the work area method takes a gradual approach depending upon the work being performed.

6.2.1 Building elements and materials. The IEBC requires that all new interior wall and ceiling finishes, floor finishes, and interior trim comply with the requirements of Chapter 8 in the IBC. In essence, depending upon the use of the space, these interior finishes may need to meet the minimum flame spread and smoke developed indexes listed in the code, as well as meet other limitations.

All windows that are replaced in R-2 and R-3 occupancies, as well as one- and two-family residences or townhomes regulated by the IRC, must comply with the requirements of Sections 702.4 and 702.5 of the IEBC. These sections are duplicates of what is required in Sections 505.2 and 505.3 of the IEBC under the prescriptive approach. They require replacement windows to include opening control devices that comply with ASTM F2090 under certain conditions and that replacement windows in emergency escape and rescue openings be no less compliant than the existing window opening. These requirements are described in detail in Secs. 5.5.1 and 5.5.2 of this book.

Section 702.6 of the IEBC requires all new materials and installations to comply with the requirements outlined in the IBC, IECC, *International Mechanical Code®* (IMC®), and *International Plumbing Code®* (IPC®). This section states that the material standards, installation requirements, and penetrations must comply with the requirements of the appropriate code. Notice that the *International Fuel Gas Code®* (IFGC®) is not listed alongside the IMC and IPC as is commonly the case. Section 702.6.1 of the IEBC states that Level 1 alterations are only required to comply with the following provisions of the IFGC, rather than the entire code:

- The "General Regulations" outlined in Chapter 3 of the IFGC, except for Sections 303.7 and 306.
- The "Gas Piping Installations" requirement in Chapter 4 of the IFGC, except for Sections 401.8 and 402.3. With that said, Sections 401.8 and 402.3 must be met when the alteration causes an increased load on the system and the size of the existing piping is not sufficient.
- The "Chimney and Vents" requirements of Chapter 5.
- The "Specific Appliances" requirements of Chapter 6.

6.2.2 Reroofing. All roofing materials and methods of installation must comply with the requirements of Chapter 15 in the IBC. This includes applying a new layer of roofing (e.g., recovering) or replacing an existing roof covering. There are two exceptions to compliance with the IBC. The first is for flat roofs only. The IBC requires flat roofs to have a minimum slope of 2 percent, or a ¼ in. rise for every run of 12 in. Section 705.1 of the IEBC does not require existing flat roofs to meet this minimum slope requirement so long as positive roof drainage is provided.

The second exception is in relation to the requirements for secondary drainage. Section 1502 of the IBC addresses the requirements for roof drainage in relation to new construction. Most primary roof drainage systems consist of gutters, integral roof drain piping, or scuppers. For flat roofs, as discussed above, the primary drain must be located at the lowest point of the roof.

Secondary drains, often referred to as emergency overflow drains, or scuppers are required when the building perimeter construction extends above the roof and water can be trapped on the roof surface. The purpose of the secondary drain is to provide backup support to the primary drains during large storm events or in the case that the primary drains might be blocked.

Figure 6.2 displays a combination primary roof drain inlet and secondary drain inlet that is fabricated by Jay R. Smith Manufacturing. The roof should be sloped to the primary drain, and then the secondary drain must be located above the primary drain a certain distance, typically this is 2 in., as shown in the figure with the secondary drain being shown on the left. It is important to note that both the primary and secondary drains are to be separately piped as required by Section 1108.2 of the IPC.

Figure 6.2 J.R. Smith 1830 combination roof drain and overflow.

Going back to the second exception to complying with the IBC Chapter 15 requirements, Section 705.1 states that secondary drains are not required for existing roofs that have positive drainage. In addition, the exception states that secondary roof drains or scuppers are not to be removed unless new drains or scuppers are added that meet the requirements of Section 1502 of the IBC.

In addition to the general reroofing requirements noted above, Level 1 alterations have specific provisions in relation to structural and construction loads, roof replacements, roof recovering, reinstallation of existing materials, and flashings when performing reroof operations. The following describes the IEBC requirements for each of these items.

Structural and construction loads. In order to perform reroofing operations, Section 705.2 of the IEBC requires that the existing roof system be capable of supporting the roof covering system as well as the material and equipment loads that will occur during the installation of the roof covering. If additional loads are added to the roof, this may require that the existing roof be evaluated to ensure that it can support the added loads. When stacking roofing materials at the roof the contractor should be careful not to overload the roof. An example would be piling materials at a cantilevered portion of the roof such as the eaves.

Roof replacements. Section 705.3 of the IEBC defines a roof replacement as removing all existing roofing materials down to the roof deck. This is often referred to as a roof tear-off. When performing a tear-off, the IEBC allows existing ice barriers to remain in place so long as an additional layer is added in accordance with Section 1507 of the IBC.

Roof recovering. Roof recovering consists of applying new roofing over an existing roofing system. This is allowed in accordance with Section 705.3.1 of the IEBC so long as the following items are met:

- The new roof covering must be installed in accordance with the manufacturer's recommendations. These recommendations must prescribe that the system can be applied over existing roofing materials and must follow the installation requirements for the specific application.
- A separate roofing system can be installed that does not rely upon the existing system for support. An example would be a standing-seam metal roof that is supported by the building's structural system rather than the roof surface.
- New protective roof coatings can be applied over select types of existing roof coverings.
- When metal panel, metal shingle, concrete tile, or clay tile roof coverings are applied over existing wood shake or wood shingle coverings, a combustible concealed space is created. Section 705.4 of the IEBC requires that this concealed space be protected by a layer of gypsum board, glass fiber, mineral fiber, or other approved noncombustible material that is secured in place.

In addition to complying with the items noted above, Section 705.3.1.1 of the IEBC does not allow roof recovers when the existing roof has two or more applications of any type of roof covering; when the existing roof covering is slate, clay, cement, or asbestos-cement tile; or when the existing roof or roof covering is water-soaked or deteriorated.

Reinstallation of materials. Section 705.5 of the IEBC allows existing slate, clay, or cement tiles to be reinstalled so long as they are not damaged. In addition, existing flashing materials, metal edges, drains, and similar items should not be reinstalled if they are rusted, damaged, or deteriorated. In no case does the IEBC allow aggregate surfacing materials used in built-up roofs to be reinstalled.

Flashings. Section 705.6 of the IEBC requires that all flashing materials comply with the requirements of manufacturer's installation instructions. In addition, metal flashing that is to receive bituminous materials must be primed prior to installation.

6.2.3 Structural. Work involving Level 1 alterations only trigger structural requirements when the alteration involves the replacement of equipment or reroofing operations. If the alterations include either of these items, then the requirements of Section 706 in the IEBC must be met. The following items describe what is required by Section 706 of the IEBC.

Gravity trigger. If the reroofing operations or equipment replacement triggers the **5 percent rule**, the existing gravity members must be evaluated to ensure compliance with the dead, live, and snow load requirements of the IBC. As described in Sec. 3.3.2 of this book, this includes any increase in snow drift loads.

This trigger does not apply to Group R occupancies that have five or fewer dwelling or sleeping units and which are constructed in accordance with the conventional construction provisions of the IBC or IRC. In addition, this trigger does not apply when a second layer of roofing is being applied over a single layer of roofing and the new roofing layer weighs no more than 3 lb/ft^2.

Unreinforced masonry (URM) parapets. Just as required when following the prescriptive path, Section 706.3.1 of the IEBC requires parapets of URM buildings to be braced. This is only required when reroofing operations are performed on at least 25 percent of the roof surface and the project is located in Seismic Design Category D, E, or F. More information in relation to the bracing of parapets is outlined in Sec. 5.3.5 of this book.

High-wind regions. When reroofing operations are being performed in high-wind regions, the existing roof diaphragm and associated connections must be evaluated in accordance with Section 706.3.2 of the IEBC. This is only required when the ultimate design wind speed is more than 115 mi/h, and a roof tear-off (e.g., roof replacement) is occurring to at least 50 percent of the roof area. This is also required when following the prescriptive approach and is described in detail in Sec. 5.3.11 of this book.

6.3 *Alterations: Level 2*

Level 2 alterations are more involved than Level 1 alterations. Section 603.2 of the IEBC describes Level 2 alterations as spaces that are reconfigured, the addition of or elimination of doors or windows, the reconfiguration or extension of any system (e.g., mechanical, electrical, plumbing, etc.), and the installation of any new equipment. As highlighted in Fig. 6.1, Level 2 alterations include not only those items that are listed in Chapter 8 of the IEBC but also those items required for Level 1 alterations as listed in Chapter 7. Thus, all the items described in Sec. 6.2 of this chapter apply to Level 2 alterations in addition to the items described below.

6.3.1 Building elements and materials. In general, Section 801.3 of the IEBC requires that all new elements and associated components conform to the requirements of the IBC. While there are a few exceptions to this requirement, they are minimal.

Existing vertical openings must conform to the requirements outlined in Section 802.2 of the IEBC. The following text describes what is required by the IEBC in relation to existing vertical openings.

Default enclosure requirements. Existing vertical openings that connect two or more floors must be enclosed. The enclosure must have a minimum 1-h fire-resistance rating in addition to approved opening protectives. This may seem like a very difficult requirement, yet there are 14 specific exceptions that would not require the owner to enclose the existing opening. In many cases, the building owner will have the option of adding fire sprinklers where none are provided or enclosing the vertical openings. The following is a brief description of the exceptions to enclosure of vertical openings:

- If enclosure of the vertical opening is not required by the IBC or *International Fire Code*® (IFC®).
- Rather than continuing the 1-h fire-resistance rating through the floor and ceiling space, blocking consisting of 2-in. nominal lumber can be used. This is not allowed at stairways.
- Enclosure is not required where the main floor connects to a mezzanine.
- Enclosure is not required if considered to be a low-hazard occupancy, the building is protected throughout by a fire sprinkler system, does not include a basement, and the main level egress capacity is sufficient to provide simultaneous egress to all building occupants.
- A 30-min enclosure is allowed for **Group A** occupancies that do not exceed three stories.
- A 30-min enclosure is allowed for **Group B** occupancies that do not exceed three stories. Such enclosure is not required if the building does not exceed 3000 ft^2 per floor level or the building is fully sprinklered.
- Enclosure is not required for **Group E** occupancies that do not exceed three stories and the building is fully sprinklered.
- Enclosure is not required for **Group F** occupancies if they do not exceed three stories, if it is for manufacturing operations and at least one stairway is protected, or the building is fully sprinklered.
- Enclosure is not required for **Group H** occupancies that do not exceed three stories and the opening is required for manufacturing purposes. In this instance each floor level must provide direct access to at least two enclosed stairways.
- A 30-min enclosure is allowed for **Group M** occupancies that do not exceed three stories. Such enclosure is not required if the opening connects no more than two floor levels or if the building is protected by fire sprinklers.
- Enclosure is not required for **Group R-1** occupancies that do not exceed three stories, and the building is either fully sprinklered or the building has fewer than 25 dwelling units above the second floor and direct access is provided to at least two approved exits.
- A 30-min enclosure is allowed for **Group R-2** occupancies that do not exceed three stories. No enclosure is required under the following conditions:
    - The vertical openings do not exceed two stories and no more than four dwelling units occur on each floor.

 - ○ The building is protected throughout by a fire sprinkler system.
 - ○ The building has no more than four dwelling units per floor, all upper story levels have direct access to at least two approved exits, and a fire alarm system is provided throughout.

- Enclosure is not required in **one- and two-family dwellings.**
- Enclosure is not required for **Group S** occupancies where the vertical opening connects no more than two floors or when it connects three floors and the building is fully sprinklered.
- Enclosure is not required for **Group S** occupancies when the IBC does not require enclosure for open parking garages or ramps.

Supplemental enclosure requirements. When the work area exceeds 50 percent of the floor area, additional enclosure provisions must be applied to existing vertical openings. It should be noted that the 50 percent requirement is for a single floor area and does not need to be 50 percent of the aggregate area of the building. The following are the two supplemental requirements that are triggered when the alteration affects more than 50 percent of the floor area:

- *Non-Stairway Openings:* Section 802.2.2 of the IEBC requires that all non-stair shafts and other floor openings are enclosed. The only exception to this requirement is if the vertical opening occurs within an individual dwelling unit.
- *Stairway Openings:* Section 802.2.3 of the IEBC requires all stairways that are part of the means of egress be enclosed with smoke-tight construction from the level of exit discharge to the highest floor where alterations occur. This enclosure requirement is only required for stairways that serve the work area. The only exception to enclosure is if the IBC or IFC does not require the stairway to be enclosed.

In addition to the requirements for enclosure of vertical openings, Section 802 of the IEBC includes provisions for smoke compartments, interior finishes, guards, and fire-resistance ratings. The following outlines the IEBC requirements for each of these items.

Smoke compartments. Where the work area occurs within a Group I-2 floor level that provides sleeping rooms to more than 30 patients, the floor level must be divided into no fewer than two separate smoke compartments. The smoke compartments are to be separated from one another by means of smoke barriers that are constructed in accordance with Section 407.5 of the IBC.

Interior finish. The interior finish applied to walls and ceilings in exits and corridors must meet the requirements of the IBC. Figure 6.3 shows what is required by Table 803.13 of the IBC. As an example, the wall and ceiling finish applied to corridors within an apartment building (Group R-2) is required to meet the Class C finish rating. If the work area exceeds 50 percent of the floor area, these interior finish requirements must be met for the entire floor if the exit or corridor serves the work area.

Guards. If not currently provided, a guard shall be installed along all portions of floors that are 30-in. or more above the floor or grade below. In addition, if existing guards are in danger of collapse, they shall be replaced. All new or replaced guards must comply with the requirements of the IBC. It is important to note that existing guards that do not meet the current IBC requirements are not required to be upgraded to comply with the IBC.

Fire-resistance ratings. There may be instances where the design professional desires to reduce the existing fire-resistance ratings of existing buildings. This can be allowed in certain instances. Section 802.6 of the IEBC allows this to occur when an automatic sprinkler system is

INTERIOR WALL AND CEILING FINISH REQUIREMENTS BY OCCUPANCY[k]						
	SPRINKLERED[l]			NONSPRINKLERED		
GROUP	Interior exit stairways and ramps and exit passageways[a,b]	Corridors and enclosure for exit access stairways and ramps	Rooms and enclosed spaces[c]	Interior exit stairways and ramps and exit passageways[a, b]	Corridors and enclosure for exit access stairways and ramps	Rooms and enclosed spaces[c]
A-1 & A-2	B	B	C	A	A[d]	B[e]
A-3[f], A-4, A-5	B	B	C	A	A[d]	C
B, E, M, R-1	B	C[m]	C	A	B	C
R-4	B	C	C	A	B	B
F	C	C	C	B	C	C
H	B	B	C[g]	A	A	B
I-1	B	C	C	A	B	B
I-2	B	B	B[h, i]	A	A	B
I-3	A	A[j]	C	A	A	B
I-4	B	B	B[h, i]	A	A	B
R-2	C	C	C	B	B	C
R-3	C	C	C	C	C	C
S	C	C	C	B	B	C
U	No restrictions			No restrictions		

For SI: 1 inch = 25.4 mm, 1 square foot = 0.0929 m^2.

a. Class C interior finish materials shall be permitted for wainscotting or paneling of not more than 1,000 square feet of applied surface area in the grade lobby where applied directly to a noncombustible base or over furring strips applied to a noncombustible base and fireblocked as required by Section 803.15.1.

b. In other than Group I-3 occupancies in buildings less than three stories above grade plane, Class B interior finish for nonsprinklered buildings and Class C interior finish for sprinklered buildings shall be permitted in interior exit stairways and ramps.

c. Requirements for rooms and enclosed spaces shall be based on spaces enclosed by partitions. Where a fire-resistance rating is required for structural elements, the enclosing partitions shall extend from the floor to the ceiling. Partitions that do not comply with this shall be considered to be enclosing spaces and the rooms or spaces on both sides shall be considered to be one room or space. In determining the applicable requirements for rooms and enclosed spaces, the specific occupancy thereof shall be the governing factor regardless of the group classification of the building or structure.

d. Lobby areas in Group A-1, A-2 and A-3 occupancies shall be not less than Class B materials.

e. Class C interior finish materials shall be permitted in places of assembly with an occupant load of 300 persons or less.

f. For places of religious worship, wood used for ornamental purposes, trusses, paneling or chancel furnishing shall be permitted.

g. Class B material is required where the building exceeds two stories.

h. Class C interior finish materials shall be permitted in administrative spaces.

i. Class C interior finish materials shall be permitted in rooms with a capacity of four persons or less.

j. Class B materials shall be permitted as wainscotting extending not more than 48 inches above the finished floor in corridors and exit access stairways and ramps.

k. Finish materials as provided for in other sections of this code.

l. Applies when protected by an automatic sprinkler system installed in accordance with Section 903.3.1.1 or 903.3.1.2.

m. Corridors in ambulatory care facilities shall be provided with Class A or B materials.

Figure 6.3 Table 803.13 of the IBC. (*Excerpted from the 2018 International Building Code; copyright 2017. Washington, D.C.: International Code Council.*)

added to the building (not when a system already exists), when the new fire-resistance ratings comply with the requirements of the current IBC, and a detailed evaluation report is provided to the code official.

The evaluation report must note if any special construction features might affect the fire-resistance ratings such as smoke-resistive assemblies, conditions of occupancy, means of egress conditions, fire protection systems, equipment, fire code deficiencies, approved modifications, or approved alternate means and methods. It is important to note that this IEBC code section is only allowed when approved by the code official.

To understand when this code section might be triggered, consider an existing nonsprinklered concrete frame building with URM infill walls. The concrete frame building meets the requirements for Type II-B construction, and Table 601 of the IBC would therefore not require a fire-resistance rating of the building frame.

The design professional desires to remove the nonstructural URM infill walls, yet these walls have an inherent fire-resistance rating, as can be calculated in Section 722 of the IBC. If the owner is willing to provide an automatic sprinkler system throughout the building, and the design professional provides a detailed evaluation, they could likely remove the URM infill walls if allowed by the code official.

6.3.2 Fire protection. Section 803 of the IEBC includes multiple triggers that require fire protection features to be provided within the altered space. When such a trigger has been met, the fire protection feature must be provided throughout the entire floor and not simply in the altered space. The fire protection triggers for Level 2 alterations can be divided into three categories: (1) fire sprinklers, (2) standpipes, and (3) fire alarms. The following text describes what is required by the IEBC for each of these three categories.

Automatic sprinkler systems. Section 803.2 of the IEBC requires fire sprinkler systems to be installed in existing buildings under two circumstances. The first is when work areas have exits or corridors that serve more than one tenant or when they serve an occupant load of greater than 30. The second circumstance is for work areas that are located on floors without openings (e.g., windowless stories).

The requirements for floors without openings are the easiest to describe. Section 803.2.3 of the IEBC requires work areas that occur at floors without openings to be sprinklered if the site has sufficient water supply without requiring the installation of a fire pump. Sprinklers do not need to be installed if Section 903.2.11.1 of the IBC would not require them for new construction.

Now going back to adding fire sprinklers to spaces where altered spaces include exits or corridors that either serve multiple tenants or provide egress to more than 30 occupants. This requirement is triggered in high-rise buildings as well as Groups A, B, E, F-1, H, I, M, R-1, R-2, R-4, S-1, and S-2. This is only triggered if sufficient water supply can be provided to the floor in question without the installation of a fire pump. For high-rise buildings, occupied tenant spaces that are outside the work area are exempt from the requirement.

For Groups A, B, E, F-1, H, I, M, R-1, R-2, R-4, S-1, and S-2, the added sprinkler protection is only required if the Section 903 of the IBC would require sprinkler protection for the occupancy if it were new construction. In addition, sprinkler protection is only required for these occupancies when the work area exceeds 50 percent of the floor area. If sufficient water supply is not provided to such spaces, an automatic smoke detection system must be provided throughout the occupiable spaces in lieu of fire sprinklers.

If a fire sprinkler system is added to the existing building, Section 803.2.4 of the IEBC requires that supervision be provided. Such supervision can be provided by a dedicated central station, an approved proprietary system, a jurisdictional remote station, or an approved local alarm service. Supervision is not required for NFPA® 13R systems that use a common supply, underground gate valves, or for most alternate extinguishing systems.

Standpipes. Section 803.3 of the IEBC requires standpipe systems to be installed where the work area includes exits or corridors that serve more than one tenant and the floor is more than 50-ft above the lowest level of fire department access. For new construction, Section 905.3.1 of the IBC requires standpipe systems for all floors that are more than 30-ft above the lowest level of fire department access.

New standpipe systems shall be installed in accordance with Section 905 of the IBC, and approved fire department connections are to be provided at each floor level. The IEBC does not require standpipe connections at floor levels that are below the lowest level of fire department access, as would be required by the IBC for new construction. The new standpipes must be capable of accepting 500 gal/min at 65 lb/in.^{2} from a fire department apparatus, or a fire pump shall be installed that supplies the riser. For sprinklered buildings, the standpipe is only required to accept 250 gal/min at 65 lb/in.^{2} from the fire department apparatus.

Section 905 of the IBC requires multiple standpipe risers to be interconnected in accordance with NFPA 14, while Section 803.3 of the IEBC does not require such interconnection when a new riser is added. If not provided, signage should be provided next to the new standpipe riser clearly noting that it is not interconnected with other standpipe risers.

Fire alarm and detection. If the work area occurs within an existing Group E, I-1, I-2, I-3, R-1, and R-2 occupancy, a new fire alarm system shall be installed if required by the IFC for new construction. Such systems shall be installed in accordance with NFPA 72, and all detection, activation, and notification devices must be listed and approved. If existing notification appliances are provided, they shall be interconnected with the new system in such a way that they will be activated throughout the building. New fire alarm systems are not required where an existing fire alarm system exists.

If the work area exceeds 50 percent of the floor area, the new fire alarm system must be provided throughout the entire floor. With that said, initiating and notification devices are not required to be added within existing tenant spaces outside of the work area.

In addition to the fire alarm requirements, Section 803.4.3 of the IEBC requires smoke alarms to be added to Group R and I-1 occupancies. This is only required if such occupancies have individual dwelling or sleeping units, and such alarms shall be located as required by the IFC. New smoke alarms within the work area shall be interconnected, but those outside of the work area do not require interconnection.

6.3.3 Carbon monoxide detection. Just as required by the prescriptive compliance method, Section 804 of the IEBC requires that carbon monoxide alarms be added anytime work occurs within a Group I-1, I-2, I-4, or R occupancy. The only time this is not required is if the work involved only affects exterior building surfaces or alterations made to plumbing or mechanical systems.

6.3.4 Means of egress. Level 2 alterations trigger multiple means of egress requirements, but it is important to note that these requirements are only triggered if the work area includes exits or corridors that serve more than one tenant. In addition, when triggered, the requirements must be met throughout the entire floor that includes the work area. In some instances, the trigger may require additional requirements beyond the floor in which the work area occurs.

Before getting into the IEBC means of egress requirements, the owner, design professional, and code official should be aware of two exceptions that can be used at any time. The first states that the requirements of Section 805 do not need to be met if it can be shown that the altered space meets the means of egress requirements of NFPA 101. The second exception states that the code official can allow the means of egress system to comply with the code in place at the time of original construction so long as he, or she, feels that it does not pose a distinct life safety hazard. While it is not likely that either of these exceptions will be used, if they are sufficient, information should be provided to the code official to substantiate their use.

The main means of egress requirements for Level 2 alterations address the required number of exits, egress doorways, openings into corridors, dead-end corridors, means-of-egress lighting, exit signs, handrails, refuge areas, and guards. The following text describes what the requirements are for each of these means of egress triggers.

Number of exits. The minimum number of exits shall be as required by Section 1006 of the IBC based upon the specific occupancy and the occupant load. A single exit is allowed by the IEBC in three separate circumstances. The first is when the travel distance and occupant load (or number of dwelling units) do not exceed what is listed in Tables 805.3.1.1 (1 and 2). These IEBC tables are shown in Fig. 6.4 and only apply to Group R-2, B, F-2, or S-2 occupancies.

STORIES WITH ONE EXIT OR ACCESS TO ONE EXIT FOR R-2 OCCUPANCIES			
STORY	OCCUPANCY	MAXIMUM NUMBER OF DWELLING UNITS	MAXIMUM EXIT ACCESS TRAVEL DISTANCE (feet)
Basement, first or second story above grade plane	R-2[a]	4 dwelling units	50
Third story above grade plane and higher	NP	NA	NA

For SI: 1 foot = 304.8 mm.
NP = Not Permitted.
NA = Not Applicable.
a. Group R-2, nonsprinklered and provided with emergency escape and rescue openings in accordance with Section 1030 of the *International Building Code*.

(a)

STORIES WITH ONE EXIT OR ACCESS TO ONE EXIT FOR OTHER OCCUPANCIES			
STORY	OCCUPANCY	MAXIMUM OCCUPANT LOAD PER STORY	MAXIMUM EXIT ACCESS TRAVEL DISTANCE (feet)
First story above or below grade plane	B, F-2, S-2[a]	35	75
Second story above grade plane	B, F-2, S-2[a]	35	75
Third story above grade plane and higher	NP	NA	NA

For SI: 1 foot = 304.8 mm. NP = Not Permitted.
NA = Not Applicable.
a. The length of exit access travel distance in a Group S-2 open parking garage shall be not more than 100 feet.

(b)

Figure 6.4 **(*a*) Table 805.3.1.1(1) and (*b*) Table 805.3.1.1(2) of the IEBC. (*Excerpted from the 2018 International Existing Building Code; copyright 2017. Washington, D.C.: International Code Council.*)**

The second circumstance in which one exit is allowed is when the work area involves the exit from a dwelling or sleeping unit in an unsprinklered Group R-1 or Group R-2 occupancy. In this circumstance, a single exit is allowed if the occupant load within the dwelling or sleeping unit is no more than 10 and the travel distance within the unit does not exceed 75 ft, or when the building is no more than three stories in height and the travel distance from the exit to any habitable room is no more than 50 ft.

The third instance in which a single exit is allowed involves Group R-2 occupancies that have no more than four dwelling units per floor. These units must be served by either an exterior exit stairway or by an interior exit stairway having a smokeproof enclosure. A single exit is only allowed if the exit access travel distance from the dwelling unit entrance to the exit is not more than 20 ft.

If the single exit requirements have not been met and a second exit is required from an existing space, Section 805.3.1.2 of the IEBC allows a new fire escape to be constructed to serve as one of the required means of egress. The provisions included in this section are very similar to what is described in the prescriptive provisions under Section 504 of the IEBC with a few additional clarifications.

If a fire escape is to be used as a means of egress component, Section 805.3.1.2.1 states that access cannot be located in a locked room so that occupants can have unobstructed access to the escape. It also clarifies that if access is through a window the minimum clear opening requirements must meet the dimensional requirements for emergency escape and rescue openings. It is important to note that new fire escapes can only be constructed when an exterior stairway cannot be constructed due to lot line restrictions.

Section 805.3.2 of the IEBC describes two instances when a mezzanine would require a minimum of two exits. Two exits are required when the occupant load of the mezzanine exceeds 50, or if the travel distance to an exit exceeds 75 ft. A 100-ft travel distance is allowed if the building is fully sprinklered.

The last item for Level 2 alterations that addresses the number of required exits is in relation to the main entrance of Group A occupancies. Section 805.3.3 requires the main exit of Group A occupancies to have an egress capacity equal to at least half the total occupant load. This is only required when the total occupant load exceeds 300. The only exception to this requirement is if the main entrance is not well defined and there are several exits evenly distributed throughout the perimeter of the building.

To help understand the main entrance requirement, consider a large theater. In most cases, the occupants all enter through the main entrance due to ticketing and access restrictions. The occupants inherently want to exit the building just as they entered. If an emergency situation occurs and everyone tries to exit at the same time, they will likely all be headed for the main entrance. As such, it is important that the code official ensure that the main entrance of existing Group A occupancies that have an occupant load of 300 or more are able to meet this requirement, or that the Level 2 alterations include provisions to increase the egress capacity of the main entrance.

Egress doorways. A minimum of two egress doorways must be provided in Level 2 alteration work areas when the occupant load exceeds 50 or when the travel distance is greater than 75 ft. The only exception to this requirement is storage rooms that have an occupant load of less than 10 or if the work area is served by a single exit as allowed by Section 805.3.1.1 of the IEBC, as discussed above.

Two egress doorways are also required for Group I-2 patient sleeping rooms or suites that exceed 1,000 ft^2 and are located within the Level 2 alteration work area. Rooms of this size likely include multiple patients and would require additional egress capacity to assist the patients in egressing the space.

In addition to requiring two egress doorways as noted above, Section 805.4 of the IEBC requires egress doors that serve 50 or more occupants to swing in the direction of egress travel and that doors entering an egress passageway or exit stairway be equipped with self-closing devices. Egress doorways serving Group A occupancies having an occupant load of greater than 100 shall be equipped with panic hardware, while power-operated sliding doors in Group I-3 occupancies shall be provided with emergency power.

Openings into corridors. Corridors often serve as a very important component in the means of egress system within the building. As such, unless the IBC would not require the corridor to be fire-resistance rated, Section 805.5 of the IEBC limits what openings can be allowed within corridors that are included in the work area. The following is a listing of opening limitations within rated corridors:

- Doors shall not be constructed of hollow core wood.
- Doors shall not have louvers.
- Group R-1, R-2, and I-1 dwelling or sleeping unit doors into the corridor must be no less than 1⅜-in. solid core or approved equivalent (e.g., 20-min fire-resistance rating) and they must be equipped with automatic door closers. Only approved glazing materials in metal frames are permitted within the door.
- Transoms in Group R-1, R-2, I-1, or I-2 occupancies shall consist of ¼-in. wired glass or fire-resistance-rated glazing.
- Openings within the corridor (e.g., sash, grille, etc.) that do not open to the outside air shall be sealed with materials that will maintain the corridor fire- and smoke-resistance ratings.

Existing doors do not need to meet the requirements noted above under two circumstances. If the building is sprinklered, the doors simply need to be tight-fitting to limit the passage of smoke and cannot contact louvers. The second way out is to show that the existing doors meet the requirements included in "Resource A" of the IEBC. This resource is entitled "Guidelines on Fire Ratings of Archaic Materials and Assemblies" and provides a guideline to the design professional and code official for evaluating the fire-resistance rating of existing building materials. This may be a common method of compliance for doors in historic buildings.

Dead-end corridors. Unless permitted by the IBC, work areas shall not include dead-end corridors that exceed 35 ft in length. If the building is sprinklered, the length of existing dead-end corridors can be by 70 ft for all but Group A and H occupancies. If newly constructed dead-end corridors will be provided and the building is sprinklered, the corridor length can be 50 ft in all but Group A and H occupancies.

Means of egress lighting. Means of egress lighting complying with the provisions of the IBC must be provided within all work areas. If the work area exceeds 50 percent of the floor area, this lighting must be provided throughout the entire floor.

Exit signs. Exit signs complying with the provisions of the IBC must be provided within all work areas. If the work area exceeds 50 percent of the floor area, these signs must be provided throughout the entire floor.

Handrails. All exit stairways that are part of the means of egress for a work area must have no less than one handrail for the full length of the exit stairway. If the existing handrail is in danger of collapse, it shall be replaced, and if the stairway is wider than 66 in., a handrail must be provided on each side. The handrails must meet the requirements of the IBC.

Refuge areas. If the alterations being performed affect existing refuge areas, the refuge area cannot be reduced below what is required by the IBC for Group I-2, I-3, and ambulatory care occupancies or as required by Section 1026.4 of the IBC for horizontal exits.

Guards. Guards must be provided along portions of stairways, landings, or balconies where located 30 in. or more above the floor or grade below. Existing guards that are in danger of collapse must be replaced. All new or replacement guards must meet the requirements of the IBC. These guard provisions apply to not only the work area in question but also to all portions of the egress path down to the level of exit discharge.

6.3.5 Structural. There are three specific structural triggers outlined in Section 806 of the IEBC. The first follows the **5 percent rule**, requiring that a structural evaluation of existing gravity elements be provided if the 5 percent trigger has been met. This trigger is for structural elements that either support additional gravity loads or that have been reduced in capacity. The same exceptions that are listed for Level 1 alterations and are described in Sec. 6.2.3 of this chapter apply to gravity members included in Level 2 alterations.

The second trigger is in relation to the lateral force-resisting-system (LFRS) within the existing building. This trigger follows the **10 percent rule** and requires an evaluation of the existing LFRS when the lateral loads have been increased by greater than 10 percent or when the structural members have been reduced in capacity by as much as 10 percent because of the alteration. Section 806.3 also notes that a seismic evaluation is required if the alteration creates a structural irregularity within the building as defined in Chapter 12 of ASCE 7-16. If an evaluation is required, it can be performed in accordance with the **reduced seismic** requirements outlined in Chapter 3 of the IEBC.

The third structural trigger listed in Section 806 is for buildings undergoing a voluntary upgrade of the LFRS. Section 806.4 of the IEBC states that the voluntary improvements that are made do not need to comply with the wind and seismic provisions of the IBC, but an analysis must still be provided to the code official. This analysis must show that the capacity of existing structural systems has not been reduced, that a new structural irregularity has not been created, that new structural members and their connections comply with the current IBC requirements, and that any new or relocated nonstructural elements are detailed and attached to the building as required by the IBC.

6.3.6 Electrical. In general, all newly installed electrical equipment and wiring must comply with the applicable provisions of the *National Electrical Code*® (NEC®). In addition, for alterations made within Group A-1, A-2, A-5, H, and I occupancies, the existing wiring must be upgraded to comply with the current requirements of the NEC.

Rather than comply with all of the current NEC requirements, Section 807.3 of the IEBC provides less restrictive requirements for Group R-2, R-3, and R-4 occupancies and for buildings that are regulated by the IRC. The following is a description of the triggers that occur within these types of dwelling units:

- *Enclosed Areas:* Enclosed areas shall have no fewer than two duplex receptacle outlets or one duplex receptacle outlet and one ceiling or wall-type lighting outlet. This is not required for closets, kitchens, basements, garages, hallways, laundry areas, utility areas, storage areas, or bathrooms.
- *Kitchens:* Kitchen areas shall have no fewer than two duplex receptacle outlets.
- *Laundry Areas:* Laundry areas shall have no fewer than one duplex receptacle outlet located near the laundry equipment. This outlet must be provided on an independent circuit.
- *Ground Fault Circuit Interruption (GFCI):* Where required by the NEC, all newly installed receptacle outlets shall be provided with GFCI protection.
- *Lighting Outlets:* Bathrooms, hallways, stairways, attached garages, detached garages with electrical power, and outdoor entrances and exits shall be provided with at least one lighting outlet.
- *Utility Rooms and Basements:* At least one lighting outlet must be provided in utility rooms or basements that are used for storage or contain mechanical equipment.
- *Equipment Clearance:* Electrical equipment must be provided with sufficient clearance as outlined in the NEC.

6.3.7 Mechanical. Either natural or mechanical ventilation must be provided in spaces that have been reconfigured and are intended to be occupied, or for areas that have been converted to habitable space. Such ventilation shall be provided as required by the IMC. If an existing mechanical ventilation system is altered, reconfigured, or extended, it shall be shown that it is able to provide not less than 5 ft^3/min of outdoor air per person and not less than 15 ft^3/min of overall ventilation air per person.

Local exhaust must be provided if the alteration will introduce equipment or processes that would produce airborne particulate matter (e.g., dust, dirt, smoke, etc.), odors, fumes, vapor, combustion products, gaseous contaminants, pathogenic or allergenic organisms, and microbial contaminants. The only exception to this requirement is if the quantity of such contaminants would not adversely affect, cause discomfort to, or impair the health of the building occupants. It should also be highlighted that the IFC has several additional requirements beyond local exhaust when several of these items are introduced within a building.

6.3.8 Plumbing. There is no plumbing trigger for Level 2 alterations if there is not an increase in occupant load. With that said, if the occupant load within a single story is increased by more than 20 percent, Section 809.1 of the IEBC requires that the quantity of plumbing fixtures provided conform to the requirements of the current IPC based upon the increased occupant load.

6.3.9 Energy conservation. As with Level 1 alterations, Section 810 of the IEBC requires that the alterations themselves comply with the requirements of either the IECC or IRC for new construction, but does not require the entire building to comply with the current energy conservation requirements.

6.4 *Alterations: Level 3*

Level 3 alterations tend to be much more involved than either Level 2 or Level 1 alterations. These consist of alterations that affect 50 percent or more of the aggregate area of the existing building. As highlighted in Fig. 6.1, Level 3 alterations include not only those items that are listed in Chapter 9 of the IEBC but also those items required for Level 1 and Level 2 alterations by Chapters 7 and 8. It is also important to note that while Level 2 alterations triggered building material, fire protection, and carbon monoxide requirements (e.g., Sections 802, 803, and 804 of the IEBC) when the work area included exits or corridors that serve more than one tenant, these requirements must be met within all Level 3 alteration work areas regardless of the number of tenants served.

The following subsections describe what the specific requirements are for Level 3 alterations beyond what is specifically required by Chapters 7 and 8.

6.4.1 Special use and occupancy. There are a few additional requirements for Level 3 alterations that are performed within existing high-rise buildings and to existing boiler and furnace rooms. When performing work within a high-rise building, the two items to pay special attention to are the existing mechanical air distribution systems as well as elevators.

If the existing recirculating air or exhaust systems have a capacity of more than 15,000 ft^3/min, they must be equipped with a smoke or heat detection device as required by the IMC. This device must be equipped with an auxiliary relay that immediately cuts off power to the fan motor before smoke is distributed to occupied spaces.

In relation to existing elevators in high-rise buildings, at least one elevator that serves public spaces and that reaches a distance of at least 25 ft above the main floor shall serve the needs of emergency personnel. This elevator must provide emergency operation as required by ASME A17.3. New elevators must include both Phase I emergency recall as well as Phase II emergency in-car operation in accordance with ASME A17.1.

Existing furnace or boiler rooms that are adjacent to Group I-1, I-2, I-4, R-1, R-2, and R-4 occupancies must be enclosed by 1-h fire-resistance-rated construction. This is not required for furnace rooms that are sprinklered or for steam boiler equipment that operates at pressures of less than 15 $lb/in.^2$ or for hot water boilers operating at a pressure of 170 $lb/in.^2$

6.4.2 Building elements and materials. Existing stairways that are open to two or more floors and which serve as a part of the means of egress shall be enclosed with 1-h fire-resistance-rated construction and approved opening protectives. Such enclosure must be provided from the floor of the highest work area down to the level of exit discharge. The stairway must also be enclosed at all floors below the level of exit discharge.

Where the work area occurs within a Group R-3 dwelling unit or affects multiple single-family dwellings (e.g., townhomes), any discontinuities in the walls separating the units shall be addressed. As an example, both the IRC and IBC require new dwelling unit separation walls to be continuous from the foundation to the underside of the roof sheathing and further clarify that such separations shall be continuous through concealed spaces.

To illustrate when a discontinuity might occur in existing fire separation walls, consider a single-story duplex that has a crawl space. If the fire-resistance rating is carried to the floor itself but does not continue to the foundation, there are two options that are available to provide continuity. The owner can elect to modify the wall by providing fire-resistance-rated construction down through the crawl space, or they can ensure that the elevated floor itself meets or exceeds the fire-resistance rating of the wall.

6.4.3 Fire protection. In addition to the triggers requiring fire sprinklers in Chapter 8, there are five additional triggers provided in Section 904.1 of the IEBC. The first requires that fire sprinklers be added to work areas in high-rise buildings so long as enough municipal water exists at the site.

The second trigger is in relation to rubbish or linen chutes. Section 904.1.2 states that sprinklers or an approved fire-extinguishing system shall be added if the IBC would require such protection. Just to clarify, Section 903.2.11.2 of the IBC requires sprinkler protection at the top of, with the terminal room, and at alternate floors for all rubbish or linen chutes. Such sprinkler protection is typically integrated into prefabricated chutes used in new construction, but many existing chutes will need to be modified.

The third fire sprinkler trigger is for occupancies that include large quantities of upholstered furniture or mattresses. These types of occupancies have large quantities of combustible materials and must be provided with sprinkler protection throughout that comply with the IBC. This requirement is only triggered for Group F-1 and S-1 occupancies exceeding 2500 ft^2 and Group M occupancies exceeding 5000 ft^2.

The fourth trigger is met if the work area affects a special occupancy specifically highlighted in the IBC. Figure 6.5 lists the spaces within a building where this requirement could be triggered. These spaces could include areas such as theatrical stages, aircraft hangars, indoor children's play areas, and much more.

The fifth, and final, trigger requiring sprinkler protection applies to work areas that have exits or corridors that serve more than one tenant and which serve an occupant load of more than 30. Such sprinkler protection does not need to be added if the IBC would not require it for new construction or if adequate municipal water is not available.

In addition to adding fire sprinklers, Section 904.2 of the IEBC requires that fire alarm systems be provided in work areas for which Section 907 of the IBC would require it for new construction. Where required by the IBC, manual alarm systems shall be provided and notification appliances shall be provided throughout the floor, which are automatically activated. Visual notification appliances are not required unless a new fire alarm system is installed or the existing system is upgraded.

ADDITIONAL REQUIRED SUPPRESSION SYSTEMS	
SECTION	**SUBJECT**
402.5, 402.6.2	Covered and open mall buildings
403.3	High-rise buildings
404.3	Atriums
405.3	Underground structures
407.7	Group I-2
410.6	Stages
411.3	Special amusement buildings
412.2.4	Airport traffic control towers
412.3.6, 412.3.6.1, 412.5.6	Aircraft hangars
415.11.11	Group H-5 HPM exhaust ducts
416.5	Flammable finishes
417.4	Drying rooms
419.5	Live/work units
424.3	Children's play structures
428	Buildings containing laboratory suites
507	Unlimited area buildings
509.4	Incidental uses
1029.6.2.3	Smoke-protected assembly seating
IFC	Sprinkler system requirements as set forth in Section 903.2.11.6 of the *International Fire Code*

Figure 6.5 Table 903.2.11.6 of the IBC. (*Excerpted from the 2018 International Building Code; copyright 2017. Washington, D.C.: International Code Council.*)

6.4.4 Means of egress. In addition to the items noted in Chapter 8, Section 905 of the IEBC requires that means of egress lighting within the exit enclosure and exit signage be provided from the highest work area floor to the floor of exit discharge. Such requirements are not limited to the work area themselves, but the egress lighting in all exit enclosures and exit signs for all floors.

6.4.5 Structural. Just as discussed in Sec. 5.3.10 of this book, Section 906 of the IBC requires a complete lateral evaluation for buildings that undergo a "substantial structural alteration." This evaluation should be performed to 100 percent of the IBC wind loads as well as **reduced seismic** level. This wind and seismic evaluation is also required for any existing buildings that are located within Seismic Design Category F. As discussed in Sec. 5.3.4 of this book, this would only occur for essential facilities (e.g., Risk Category IV) located in high-seismic regions.

Level 3 alterations also trigger the following items, which were discussed in detail in Sec. 5.3 of this book when following the prescriptive path:

- *Anchorage of Concrete and Masonry Buildings:* Evaluating existing roof-to-wall and floor-to-wall anchors in existing concrete or masonry buildings that have flexible diaphragms and are in Seismic Design Category D through F. This evaluation should be performed to the **reduced seismic** level.

- *Anchorage of URM Buildings:* Roof-to-wall anchors shall be added in URM buildings located in Seismic Design Category C through F unless an evaluation performed to the **reduced seismic** level shows that such anchors are not required.
- *Anchorage of URM Parapets:* Parapet bracing shall be provided at URM parapets located in Seismic Design Category C through F unless an evaluation performed to the **reduced seismic** level shows that such bracing is not required.
- *Anchorage of URM Partitions:* Existing URM partitions that are located in Seismic Design Category C through F and which are located adjacent to an egress path shall either be anchored, removed, or altered. This work shall be accomplished to ensure that the anchored or altered URM partition is able to resist the **reduced seismic** out-of-plane forces.

6.4.6 Energy conservation. Just as with Level 1 and Level 2 alterations, Level 3 alterations do not require the entire building to conform to the energy conservation requirements of the IECC or IRC. Only those portions that are undergoing alterations are required to conform to the IECC or IRC energy requirements.

6.5 *Change of Occupancy*

As discussed earlier in this chapter, a change of occupancy can occur anytime the occupancy classification changes, the occupancy changes from one group to another within the same occupancy classification, or the space will still fall under the same occupancy group but the use triggers additional provisions within the code. Changes in occupancy should never occur unless approved by the code official, and such spaces shall not be occupied until all necessary work has been completed and a certificate of occupancy has been issued.

Just as with alterations, changes in occupancy have multiple triggers that come into play based upon use, structural, electrical, and much more. The key difference for changes in occupancy is that rather than reference the requirements of the IBC, scoping provisions for fire protection and means of egress are provided in Section 1011 of the IEBC. The following subsections will break down the change of occupancy triggers in more detail.

6.5.1 Special use and occupancy. Anytime the change of occupancy causes the existing building, or a portion of the existing building, to have one of the following special uses, the building shall be made to comply with all of the requirements of the current IBC:

- Covered and open mall buildings
- Atriums
- Motor vehicle–related occupancies
- Aircraft-related occupancies
- Motion picture projection rooms
- Stages and platforms
- Special amusement buildings
- Incidental use areas
- Hazardous materials
- Ambulatory care facilities
- Group I-2 occupancies

In addition to the above special occupancies, anytime the change of occupancy occurs within an existing underground building, the existing building must be brought into compliance with the IBC requirements for new underground buildings.

6.5.2 Structural. The structural requirements for changes in occupancy under the work area method are the same as those provided under the prescriptive approach, which were discussed in Sec. 5.6.5 of this book. The following four bullets briefly describe the four structural triggers, but the reader should review Sec. 5.6.5 to better understand these structural requirements.

- *Live Loads:* Anytime the change in occupancy causes an increase in live loads, the existing structural elements supporting these loads shall be evaluated to ensure compliance with Section 1607 of the IBC. The **5 percent rule** applies, allowing for a live load increase of no more than 5 percent without having to show the existing structural members meet the current IBC.
- *Snow and Wind Loads:* If the change of occupancy causes the building to be classified into a higher risk category, in accordance with Table 1604.5 of the IBC, the structural members must be evaluated to verify they can resist the current IBC snow and wind loads. The only exception to this requirement is if the change in use affects no more than 10 percent of the existing building.
- *Seismic Loads:* Similar to above, if the change of use causes the building to be reclassified into a higher risk category, the entire building must be shown to meet the seismic provisions of the IBC. This is not required if the building is reclassified to a Risk Category III and is in Seismic Design Category A or B, or if change affects less than 10 percent of the aggregate area of the building and it does not involve an essential facility (e.g., Risk Category IV). It should also be noted that URM buildings assigned to Risk Category III and located in Seismic Design Category A or B can use the appendix (e.g., **reduced seismic** level) to show compliance rather than complying with the **full seismic** level.
- *Access to Risk Category IV:* Any structure that provides "operational access" to an essential facility (e.g., Risk Category IV) shall be evaluated to show compliance with the snow, wind, and seismic requirements of the IBC. This only applies if the change in occupancy causes this to occur. Please review "Access to Risk Category IV" in Sec. 5.6.5 of this book.

6.5.3 Electrical. Changes in occupancy can require a significant number of electrical triggers within existing buildings. In all instances, the code official should require unsafe electrical conditions to be addressed. Also, the number of required electrical outlets within the altered spaces must comply with the current provisions of the NEC. In addition, Section 1007.3 of the IEBC requires that changes in occupancy within existing buildings undergo an electrical service upgrade to show compliance with the current NEC.

The most onerous electrical requirement in Chapter 11 requires that the electrical wiring and equipment within the building be upgraded to comply with the current NEC. This is obviously a very big deal, but it is only triggered when the change of occupancy involves one of the following special occupancies:

- Hazardous locations
- Commercial garages, repair, and storage
- Aircraft hangars
- Gasoline dispensing and service stations
- Bulk storage plants
- Spray application, dipping, and coating processes

- Health care facilities
- Places of assembly
- Theaters, audience areas of motion picture and television studios, and similar locations
- Motion picture projectors
- Agricultural buildings

6.5.4 Mechanical. Existing buildings undergoing a change of occupancy have only two mechanical triggers. The first is if the change in use causes new or altered kitchen exhaust requirements, while the second is if the space would now require increased mechanical ventilation. If either case applies, the mechanical systems in question shall be shown to comply with the current requirements of the IMC.

6.5.5 Plumbing. When a building undergoes a change of occupancy, the design professional should compare the existing number and type of plumbing fixtures to what is required in the IPC. If the IPC would require an increase in plumbing fixtures or requires different plumbing fixtures than are provided, the code official shall ensure that the "intent" of the IPC is met for the new occupancy. In addition, Section 1009.1 of the IEBC states that if the change of occupancy results in an increased water supply, the requirements of the IPC should also be met.

If the new occupancy is a food-handling establishment, all existing sanitary waste lines that occur within food or drink preparation areas or within food storage areas shall be "panned" or otherwise protected to prevent leaking. New drain lines also cannot be installed above these areas, and lines shall be protected as required by the IPC. If the change of use causes grease or oil-laden wastes to be produced, a grease interceptor shall be installed in accordance with the IPC.

Should the change of use result in an occupancy that produces chemical wastes, the piping shall be shown to be compatible with the wastes that will be produced. If not compatible, the owner will need to show the code official how the waste will be neutralized prior to entering the drainage system. In both cases, the chemical waste shall not be discharged to the public sewer unless approved by the local sewage authority.

Last of all, if the occupancy group is changed to Group I-2, the entire plumbing system shall be made to comply with the requirements of the IPC.

6.5.6 Light and ventilation. Section 1010.1 of the IEBC requires that the light and ventilation provided for the new occupancy comply with the requirements of the IBC for new construction.

6.5.7 Fire protection. In general, existing buildings that undergo a change of occupancy shall meet all the fire protection requirements listed in Chapter 9 of the IEBC for Level 3 alterations. If the change of occupancy is separated by means of fire barriers complying with the IBC, the Chapter 9 requirements only apply to the separated portion. If a separation is not provided, the entire building must comply with the Chapter 9 provisions. In addition, buildings undergoing a change of occupancy must comply with the following supplemental fire protection provisions.

Fire protection systems. If the change of occupancy causes the space to have a different fire protection system threshold, in accordance with Chapter 9 of the IBC, an automatic fire sprinkler system and/or fire alarm system must be provided as required by the IBC.

As an example, consider an existing office building that is being converted into a charter school. The building in question is two stories and has a total square footage of 12,500 ft^2. The Group B occupancy was not required to be sprinklered per Section 903 of the IBC, nor were sprinklers required to increase the allowable area in accordance with Chapter 5 of the IBC. In addition, Section 907 of the IBC did not require a fire alarm system for the building. The new

Group E occupancy, however, must be both sprinklered and alarmed in accordance with Sections 903.2.3 and 907.2.3 of the IBC. As such, fire sprinklers and fire alarms must be added throughout.

Interior finish. The interior finishes must comply with the IBC requirements for the new occupancy.

Heights and areas. Table 1011.5 of the IEBC (see Fig. 6.6) lists the hazard categories for different occupancy groups in relation to building height and area. If the change of occupancy causes the building to be classified into a higher hazard category, the existing building must comply with the allowable heights and area provisions of Chapter 5 in the IBC. If it is classified as the same or as a lesser hazard category, the existing height and area shall be deemed acceptable.

HEIGHTS AND AREAS HAZARD CATEGORIES	
RELATIVE HAZARD	OCCUPANCY CLASSIFICATIONS
1 (Highest Hazard)	H
2	A-1; A-2; A-3; A-4; I; R-1; R-2; R-4, Condition 2
3	E; F-1; S-1; M
4 (Lowest Hazard)	B; F-2; S-2; A-5; R-3; R-4, Condition 1; U

Figure 6.6 Table 1011.5 of the IEBC. (*Excerpted from the 2018 International Existing Building Code; copyright 2017. Washington, D.C.: International Code Council.*)

Meeting the requirements for higher hazard categories may be hard in several instances. For that reason, Section 1011.5.1.1 of the IEBC provides an exception that creates a win-win for the owner and the code official. This section allows the building to be separated into separate buildings by use of fire barriers rather than fire walls. This will allow the owner to use existing or new walls constructed to the lesser provisions in relation to fire barriers. As an example, it may be hard to construct a new fire wall that meets the structural stability requirements of the IBC.

There are three important items to keep in mind if this route is taken. First, the entire building must be sprinklered and an allowable area increase cannot be used due to the sprinklers. Second, each separated area must meet the nonsprinklered area limitations of the IBC. Lastly, the fire barrier must meet the fire-resistance rating for fire walls as listed in Table 706.4 of the IBC.

Exterior walls. Just as with height and area, Table 1011.6 of the IEBC (see Fig. 6.7) provides hazard categories that are in relation to exterior wall exposures. If the change of occupancy causes the building to be classified into a higher hazard category, the existing exterior walls must meet the requirements of the IBC for new construction. Section 1011.6.1 of the IEBC does have one exception to this requirement. The exception allows for exterior walls to have a 2-h fire-resistance rating if the building does not exceed three stories and it is classified as a Group A-2, A-3, B, F, M, or S occupancy. The Group A-2 and A-3 occupancies must have fewer than 300 occupants to utilize this exception.

In addition to the exterior wall fire-resistance rating, Section 1011.6.3 of the IEBC requires openings in exterior walls to meet the requirements of the IBC. This same section allows for protected openings to total as much as 50 percent of the aggregate area of the exterior walls.

EXPOSURE OF EXTERIOR WALLS HAZARD CATEGORIES	
RELATIVE HAZARD	OCCUPANCY CLASSIFICATIONS
1 (Highest Hazard)	H
2	F-1; M; S-1
3	A; B; E; I; R
4 (Lowest Hazard)	F-2; S-2; U

Figure 6.7 Table 1011.6 of the IEBC. (*Excerpted from the 2018 International Existing Building Code; copyright 2017. Washington, D.C.: International Code Council.*)

If the change of occupancy does not cause the building to be classified to a higher hazard category, then the existing exterior walls, including openings, shall be considered acceptable.

Vertical shafts. Section 1011.7.1 states that vertical shafts shall be designed in accordance with the requirements of the IBC, or they are required to meet the atrium provisions in Section 404 of the IBC. If the change of occupancy causes the building to be classified to a higher means of egress hazard category (see Fig. 6.8), interior stairways and other vertical shafts (e.g., elevator hoistway, utility shafts, etc.) shall be enclosed as required by the IBC.

MEANS OF EGRESS HAZARD CATEGORIES	
RELATIVE HAZARD	OCCUPANCY CLASSIFICATIONS
1 (Highest Hazard)	H
2	I-2; I-3; I-4
3	A; E; I-1; M; R-1; R-2; R-4, Condition 2
4	B; F-l; R-3; R-4, Condition 1; S-1
5 (Lowest Hazard)	F-2; S-2; U

Figure 6.8 Table 1011.4 of the IEBC. (*Excerpted from the 2018 International Existing Building Code; copyright 2017. Washington, D.C.: International Code Council.*)

Interior stairways are not required to be enclosed under three circumstances as outlined in Section 1011.7.2 of the IEBC. First, enclosure is not required for all but Group I occupancies if the stairway is only serving one adjacent floor and is not connected to corridors or other stairways. Second, existing stairways do not require enclosure if 1-h fire-rated horizontal assemblies separate each floor level and at least one sprinkler is provided above each opening on the tenant side. Last, existing penetrations into the stair shaft are acceptable if they are protected as outlined in Chapter 7 of the IBC.

For other than stairways, existing vertical shafts that have a 1-h fire-resistance rating shall be deemed acceptable even if the IBC would require a higher fire-resistance rating for new construction. If the entire building is sprinklered, the occupancy does not include Group I, and vertical openings connect fewer than six stories, enclosure is not required for openings other than stairways.

6.5.8 Means of egress. Section 1011.4 of the IEBC addresses the means of egress requirements for existing buildings that undergo a change of occupancy. This section requires that, at a minimum, the egress capacity comply with the requirements of the IBC, while the handrails and guards meet the Level 2 alteration requirements of Section 805 in the IEBC. In addition, if the change of occupancy is to a higher "means of egress hazard category," the entire means of egress shall be shown to comply with the provisions of the IBC.

Table 1011.4 of the IEBC (see Fig. 6.8) notes what "means of egress hazard category" applies to each specific occupancy classification. For example, if the change of occupancy is of an office space (e.g., Group B) that is converted to a restaurant (e.g., Group A), Fig. 6.8 shows that it will go from a relative hazard 4 to a relative hazard 3. As such, this change of occupancy would require that the entire means of egress system comply with the requirements of the IBC.

There are seven exceptions in relation to full compliance with the IBC for the means of egress system. Each of those exceptions is described below:

- Stairways can be enclosed as outlined in Section 903.1 of the IEBC.
- Existing stairways, including handrails and guards, can be retained as is subject to approval from the code official.
- Stairways are not required to meet the maximum riser height and minimum tread depth where the existing pitch or slope cannot be reduced.
- Existing corridor walls shall be permitted where both sides of wood lath and plaster are in good condition or ½-in.-thick gypsum is provided.
- Existing corridor doors, transoms, and other openings shall comply with Section 805.5 of the IEBC.
- Existing dead-end corridors shall comply with Section 805.6 of the IEBC.
- Existing operable windows with a minimum openable area of 4 ft^2, height of 22 in., and width of 20 in. will be deemed acceptable for emergency escape and rescue openings.

Section 1011.4.2 of the IEBC addresses what is required if the means of egress hazard category does not change or if it is less than in the previous occupancy. If this is the case, the existing means of egress system will simply need to comply with Section 905 of the IEBC. All newly constructed means of egress components are required to comply with Chapter 10 of the IBC.

There are a few items to highlight that are not discussed in Chapter 10 of the IEBC. First, the reader likely noticed that Chapter 10 does not require existing buildings that undergo a change in use to meet the energy conservation requirements of the IECC or IRC as would be required for new construction. In addition, changes in occupancy typically trigger alterations. As such, items such as bracing a URM parapet would likely be required for buildings undergoing changes in occupancy, although this is not specifically noted in Chapter 10.

6.6 *Additions*

In all cases, the addition itself must be constructed to fully conform to the applicable requirements of the International codes. In some cases, as will be discussed in this section, the addition may also trigger the requirement for the existing building to comply with portions of the current International codes. It is also important to note that any addition cannot create or make worse a nonconformity in the existing building in relation to accessibility, structural strength, fire safety, means of egress, or the capacity of mechanical, plumbing, or electrical systems.

Besides the above, additions can cause triggers to the existing building in relation to the allowable height and area, structural requirements, smoke and CO detectors, storm shelters, and energy conservation. The specific requirements in relation to each of these items is noted below.

6.6.1 Heights and areas. The entire building, including additions, must meet the allowable height and area limitations of Chapter 5 in the IBC. The allowable area does not need to comply if a fire wall is constructed between the addition and existing building. Section 1102.2 of the IEBC also includes an exception, which allows existing floor openings or nonoccupiable appendages to be infilled or used as new usable space without meeting the allowable area limitations of the IBC. Existing fire areas that are increased because of the addition shall be made to comply with the requirements of Chapter 9 in the IBC.

6.6.2 Structural. The **5 percent gravity rule** and **10 percent lateral rule** both apply to additions. If the gravity loads (e.g., dead, live, and snow) are increased by more than 5 percent, or the existing members capacity has been reduced by 5 percent, the structural members shall be evaluated to conform to the gravity requirements of the IBC. Likewise, if the lateral loads to any existing structural elements are increased by more than 10 percent, they will need to be evaluated to show compliance with the IBC. This evaluation must be performed to the IBC **full seismic** level. Group R occupancies that have no more than five dwelling units and conform to the conventional construction requirements are exempt from the 5 percent and 10 percent rules.

As discussed throughout this chapter, the code official must pay special attention when the existing building is located within a flood hazard area. Section 1103.3 of the IEBC provides five different scenarios and lists different requirements for each when additions are made within flood hazard areas. Those scenarios and their requirements are listed below.

- *Horizontal Addition—Structurally Connected:* If the addition and any other associated work constitute "substantial improvement," both the addition and the existing building must meet the flood design requirements of the IBC or IRC.
- *Horizontal Addition—Structurally Independent:* The addition itself shall comply with the flood-resistant design requirements of the IBC or IRC. If the addition and any other associated work constitute "substantial improvement," the existing building is also required to be brought up to the flood design requirements of the IBC or IRC.
- *Vertical Addition:* If the vertical addition and any other associated work constitute "substantial improvement," the existing building is also required to be brought up to the flood design requirements of the IBC or IRC.
- *Raised or Extended Foundation:* If the existing foundation is to be raised or extended and the associated work constitute a "substantial improvement," the existing building shall be brought up to the flood design requirements of the IBC or IRC.
- *New or Replacement Foundation:* All new or replacement foundations must meet the flood design requirements of the IBC or IRC.

6.6.3 Smoke alarms. Additions made to existing Group R and I-1 occupancies require that smoke alarms be installed within the addition and throughout the existing building as required by Section 1103.8 of the IFC or Section R314 of the IRC.

6.6.4 Carbon monoxide alarms. As with smoke alarms, additions made to existing Group I-1, I-2, I-4, and R occupancies require that carbon monoxide alarms be installed within the addition and throughout the existing building as required by Section 1103.9 of the IFC or Section R315 of the IRC.

6.6.5 Storm shelters. If an addition with an occupant load of 50 or more is made to an existing Group E occupancy, and the tornado design wind speed is 250 mi/h or more, a storm shelter must be provided that complies with the requirements of ICC® 500. This trigger does not apply to Group E daycare facilities or Group E occupancies that are accessory to places of religious worship. The storm shelter shall be located within the building it serves or such that the travel distance from the building to the shelter does not exceed 1000 ft.

If a storm shelter must be constructed, it shall provide sufficient capacity for the occupants of all buildings at the site. The occupant load count should consider the combined occupant load of the classrooms and offices or of the indoor assembly spaces. If the addition itself is not large enough to accommodate the total occupant load, the storm shelter shall at least have sufficient capacity to accommodate the occupants of the addition itself.

6.6.6 Energy conservation. The addition itself is required to comply with the energy conservation requirements of the IECC or IRC. With that said, many times alterations are made to the existing building along with the additions, and those alterations will also need to comply with the IECC or IRC requirements as required by Chapters 7 to 9 of the IEBC.

6.7 *Historic Buildings*

Chapter 12 of the IEBC prescribes what is required for historic buildings under the work area approach. Section 5.7 of this book discussed the historic building requirements when following the prescriptive approach. In that section, there were three specific triggers that were addressed for historic structures and involved life safety hazards, flood hazard areas, and general structural triggers.

There are a few items in relation to the prescriptive approach that are more onerous for both the owner and for the code official than when following the work area approach. For one, the prescriptive approach puts the burden on the code official to identify life safety hazards and to have them remedied. The second item is in relation to the structural triggers, where the prescriptive approach would require the owner to upgrade the structural portion of the building to current IBC requirements. The work area method provides a more reasonable method of dealing with historic buildings for both the code official and the owner.

In order to use the provisions included in Chapter 12 of the IEBC, the building must be designated a historic structure. The definition of a historic building, as defined in Section 202 of the IEBC, is broken into three parts. It states that structures can be considered historic buildings when they are (1) listed on the National Register of Historic Places; (2) designated as historic under an applicable state or local law; or (3) certified as a contributing resource under a national, state, or local designated historic district. In either case, if allowing the provisions of this chapter to be used, the code official should require appropriate documentation from the owner showing that the building or structure meets the IEBC definition of historic.

There are seven requirements that need to be highlighted in Chapter 12 of the IEBC. Those items include the requirements for a historic building report, flood hazard areas, repairs, fire safety, change of occupancy, structural, and relocated buildings. In addition to these items, all historic buildings are required to comply with the accessibility requirements listed in Chapter 3 of the IEBC and discussed in Sec. 3.6.10 of this book. The following subsections will break down the requirements of the seven items noted above.

6.7.1 Historic building report. For the code official, the greatest benefit of the work area method versus the prescriptive method is the requirement for a historic building report. This is prescribed in Section 1201.2 of the IEBC and requires such a report for any historic structure that is undergoing an alteration or change of occupancy. This report must be prepared by a registered design professional and must identify each code-required safety feature that *is provided* within the facility. The report should also identify items that *are not* in compliance with the code and should describe how conformance to the code requirements would negatively affect the historic nature of the building.

For buildings that are located within Seismic Design Category D, E, or F, the report should include a structural evaluation that describes the vertical and horizontal members of the lateral force-resisting system (LFRS). The LFRS evaluation should describe specific strengths and weaknesses that exist.

The key to this report is that it will clearly identify all structural and other life safety features of the facility that are not in compliance with the currently adopted code. When following the prescriptive compliance path, Section 507.2 of the IEBC requires the code official to determine what items were not in compliance. While the historic building report will note specific deficiencies, that does not necessarily mean that each item will need to be remedied. The code official should work closely with the owner and design team to ensure that a reasonable approach is followed.

Regardless of what is decided between the code official and the owner, Section 1201.5 of the IEBC requires that all unsafe conditions be remedied. There will be some items that cannot meet the letter of the code, as they will affect the historic nature of the building. In this case, the code official may want something in writing from the state historic architect, or another authority, that can be kept in the project file and protect the jurisdiction should something arise regarding that item in the future.

6.7.2 Flood hazard areas. As has been stated multiple times throughout this text, if the project is located within a flood hazard area and the work being performed constitutes a substantial improvement, it shall be brought up to the current flood design requirements of the IBC or IRC. Like the prescriptive path, the work area method *does not require* historic buildings to meet the current flood design requirements when they will continue to be listed as historic structures after the alterations or changes in occupancy have been performed.

6.7.3 Repairs. Any repairs or replacement materials used within historic buildings can utilize like materials and methods of construction. There are only two exceptions to this allowance: (1) if the materials are hazardous (e.g., asbestos, lead paint, etc.), and (2) replacement glazing in hazardous locations must still meet the requirements of Chapter 24 in the IBC.

6.7.4 Fire safety. Section 1203 of the IEBC specifies the fire safety provisions for all historic buildings that are undergoing alterations, changes in occupancy, or those that are being moved. In many instances it may be hard to cause a historic building to comply with the provisions for new buildings. If specific fire safety provisions are required and they cannot be met, the owner has the option of installing an approved fire-extinguishing system in accordance with Sections 1203.2 and 1203.12 of the IEBC. Regardless of whether this exception has been used, the required number of exits must be provided from any facility.

In relation to the means of egress, Section 1203.3 of the IEBC allows the existing door openings, corridors, and stairway widths to remain as is, even if less than required by the code for new construction, provided the code official deems sufficient width is provided. In addition, the main entrance door is not required to swing in the direction of egress travel, provided other means of egress are provided that can accommodate the entire occupant load of the building.

The following bullets address other fire safety–related items that need to be addressed in historic buildings:

- *Interior Finishes:* If interior finishes are deemed "historic" they can remain and are not required to comply with the requirements of the IBC.
- *Stairway Enclosure:* Existing stairways are not required to be fire-resistance-rated, provided they incorporate elements that would not allow the passage of smoke. This includes items such as tight-fitting doors and sealed penetrations. This is only allowed for stairways serving three stories or fewer.
- *One-Hour Fire-Resistance:* Where the IEBC or IBC would require 1-h fire-resistance-rated construction, it is not required for historic buildings, provided the wall or ceiling finish consists of wood, metal lath, or plaster.

- *Glazing in Fire-Resistance-Rated Systems:* Glazing is allowed within interior walls with a 1-h fire-resistance rating so long as it is provided with approved smoke seals and the area affected is provided with automatic fire sprinklers.
- *Stairway Railings:* Grand staircases in historic buildings, such as the one shown in Fig. 6.9, do not require handrails and guards that comply with current code requirements. This is a huge difference from what is required in Section 1014 of the IBC for new construction, as handrails would need to be provided not only along the lower stairs shown in Fig. 6.9 but also at intermediate locations so that anywhere along the stairway width persons can be within 30-in. of a handrail. In addition to the grand staircase provision, existing nonconforming handrails in historic structures only need to be replaced if they are considered to be structurally dangerous (e.g., not capable of resisting the code-required 200 lb concentrated loads or 50 lb per lineal foot load).

Figure 6.9 Grand staircase at San Francisco City Hall. (*Photo Credit: Michael Creasy Photos.*)

- *Guards:* Existing guards are not required to meet the height limitations of Section 1015.4 of the IBC, nor are openings required to meet the requirements of Section 1015.4 of the IBC. Most historic buildings will have shorter guards, and balusters will allow the passage of objects much larger than 4-in. in diameter. In accordance with Section 1203.10 that is acceptable for historic buildings.
- *Exit Signs:* Section 1203.11 of the IEBC states that the IBC requirements for exit signs and egress path marking are not required in historic buildings if they would damage the historic character of the building. In this instance, alternative methods to identify exits and the egress path should be provided. Figure 6.10 portrays an alternative exit sign that is provided within the Illinois Statehouse.

Figure 6.10 Illinois statehouse exit sign. (*Mackey, 2016.*)

6.7.5 Change of occupancy. In general, historic buildings must comply with the requirements of Chapter 10 of the IEBC when undergoing a change of occupancy. Section 1204 of the IEBC outlines a few instances when the full requirements of Chapter 10 do not need to be met and lesser provisions can be complied with. Those instances are as follows:

- *Building Area:* Rather than having to meet the building area limitations of Chapter 5 of the IBC, Section 1204.2 of the IEBC allows historic buildings to exceed the Chapter 5 allowable area requirements by 20 percent.
- *Location on Property:* Historic buildings that undergo a change of occupancy, which are reclassified to a higher exterior wall hazard category (e.g., Table 1011.6 of the IEBC), may use alternative methods to comply with the exterior wall fire-resistance rating and opening protective requirements.
- *Occupancy Separation:* If 1-h occupancy separations are required by the IBC, they can be omitted, provided the building is protected throughout by an automatic fire sprinkler system. Required separations of more than 1 h are not exempt.
- *Roof Covering:* There are instances where the IBC requires roofing to consist of Class A, B, or C roof coverings. Class A roof coverings are the most effective against severe fire test exposures. Table 1505.1 of the IBC requires a minimum of a Class B roof covering for buildings of Type IA, IB, IIA, IIIA, IV, or VA construction. Section 1204.5 of the IEBC allows new roof coverings for historic buildings to simply meet the Class C fire rating, regardless of use or occupancy.
- *Means of Egress:* Existing doors, corridors, or stairway widths that do not comply with the IBC requirements for new construction can remain as is, provided the code official determines that enough egress width is provided for the space based upon the occupant load. Section 1204.6 of the IEBC states that the code official may determine that operational control limits and maximum occupancy limits be established in order to use this exception.

- *Door Swing:* Existing front doors serving historic buildings are not required to swing in the direction of egress travel. This exception is only allowed if other approved exits exist in the building that can serve the entire occupant load for the building without considering the main entrance.
- *Transoms:* Existing transoms can remain within fire-rated corridor walls provided they are fixed in the closed position and they consist of wired glass, or another form of approved glazing is provided on at least one side of the transom.
- *Finishes:* Based upon the occupancy, if the IBC would require Class C finish materials, an approved fire-retardant paint can be placed over the existing finish. This is not required if the building is sprinklered throughout and the existing finish materials are deemed to add to the historic character of the building.
- *One-Hour Fire-Resistance Assemblies:* If the IBC would require 1-h fire-resistance-rated assemblies, they are not required for historic buildings that undergo a change of occupancy, provided the existing wall and ceiling finish materials consist of wood lath and plaster.
- *Stairways and Guards:* In general, stairways and guards should comply with Section 1203 of the IEBC. There are, however, two exceptions noted in Section 1204.11 of the IEBC. The first allows the code official to grant an exception if he or she feels the existing stairways meet the intent of the code. The second exception states that existing conditions can remain "as is" if the buildings are less than 3000 ft^2.
- *Exit Signs:* Section 1204.12 of the IEBC allows the code official to accept alternative locations for exit signs. This is only allowed if the standard exit sign locations would damage the historic character of the building. Alternative signs must still clearly identify the exits and exit path.
- *Exit Stair Live Load:* The IBC requires stairways to be designed for a live load of 100 lb/ft^2. Section 1204.13 of the IEBC allows existing stairways in historic buildings that are changed to a Group R-1 or R-2 to be evaluated for a live load of 75 lb/ft^2. All other occupancies are required to be evaluated for a 100 lb/ft^2 live load, except for stairways within an R-2 occupancy.
- *Natural Light:* Section 1010 of the IEBC requires existing buildings that undergo a change of occupancy to meet the light and ventilation requirements of the IBC. For historic buildings undergoing a change of occupancy, Section 1204.14 allows the existing level of natural light to be sufficient if adding more lighting would affect the historic character of the building.

6.7.6 Structural. Section 1205.1 requires historic buildings to comply with the structural provisions outlined in Chapters 4 and 5. In other words, the structural requirements must comply with the repair or prescriptive compliance provisions of the IEBC. The reader should review Sec. 5.7.3 of this book to have a better understanding of what the structural requirements for historic buildings are under the prescriptive compliance method.

6.7.7 Relocated buildings. Section 1206 of the IEBC notes that historic buildings that are moved are to conform to the requirements of Chapter 12 of the IEBC, except that the new footings and foundations must conform to the requirements of the IBC. This section also clarifies that the building must be sited upon the new lot so that it complies with the requirements of the IBC (e.g., fire-resistive construction, accessibility provisions, etc.).

CHAPTER 7

PERFORMANCE COMPLIANCE METHOD

7.1 *Background*

The majority of code requirements are prescriptive in nature. This means that they tell the user exactly what should be done and have very limited flexibility. With that said, the design community is slowly pushing toward the concept of performance-based design (PBD). The performance approach focuses on what the building is required to do, rather than prescribing how it is to be constructed. PBD requires the design professional to work more closely with the owner of the building, as well as the code official, to ensure that the design meets specific requirements for the occupants and not simply the prescriptive requirements of most building codes.

The International codes, as well as many European codes and standards, now include several provisions for PBD. Chapter 13 of the *International Existing Building Code*® (IEBC®) provides PBD approaches for existing buildings, while a stand-alone performance code, the *ICC Performance Code for Buildings and Facilities*® (ICCPC®), has been developed that can be used for new construction as well (see Fig. 7.1). Section 101.1 of ICCPC states that the intent of this performance code is to promote an innovative, flexible, and responsive solution to best optimize both cost and materials.

Figure 7.1 ICC Performance Code for Buildings and Facilities. (*Copyright 2017. Washington, D.C.: International Code Council.*)

In an article written in the *Forest Products Journal*, Greg Foliente highlights three benefits of the PBD approach over the prescriptive requirements of the IBC (Foliente, 2000). The first benefit is *innovation*. Performance approaches can consider several different types of materials, whereas prescriptive methods define the materials that are to be used. In essence, PBD can lead to the development of new materials but also to new and better ways of analysis or construction methods.

The second benefit of PBD described by Foliente is the ability to *cost optimize*. Not only could a reduction in costs come from the use of newer and more innovative materials, as discussed in benefit 1, but PBD requires the designer to focus on the use of the space and allows them to cut costs in other areas that are not as crucial to the use of the space. A common method used for cost optimization is called "value engineering." Value engineering provides a systematic approach for either improving the function or reducing the cost (see Fig. 7.2). As shown in the figure, basic functions should not be reduced as a consequence of pursuing value improvements.

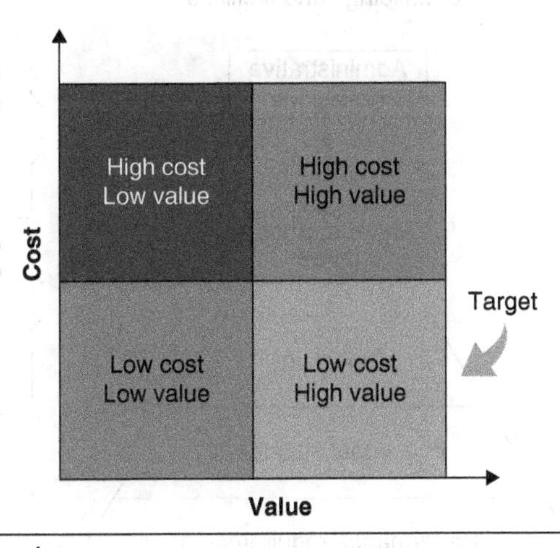

Figure 7.2 Value engineering.

The third, and final, benefit described by Foliente involves *international trade*. In his article, Foliente describes the prescriptive criteria included in many building codes as being barriers to international trade. He refers to the following statement made by the World Trade Organization (WTO), "*Wherever appropriate, members shall specify technical regulations based on product requirements in terms of performance rather than design or descriptive characteristics.*" Members of the WTO recognize the importance of performance over prescriptive requirements.

To better understand the benefits of PBD, consider the following two examples. First, a contractor chooses to meet the braced wall provisions of the IRC rather than hire an engineer to design a new single-family home. The braced wall provisions of the IRC are quite conservative and will likely lead to more walls requiring structural wood sheathing. If the home were to be engineered, it is quite likely that less sheathing and even larger window and door openings could be used on the home than would be allowed under the IRC approach.

Similarly, consider the egress requirements for a new commercial building. The IBC would prescribe the occupant load count for the building and would also specify a maximum length and minimum egress width for the occupants to exit the building. Rather than meeting the IBC-prescribed egress length and width limitations, a PBD approach could be to analyze the number of occupants within the building and ensure that they are able to exit the building within a certain amount of time. Something similar is done when designing a smoke control system within a building.

7.2 *Components of Performance-Based Design*

Figure 7.3 displays the performance pyramid that is included at the front of the ICCPC. This pyramid helps the user understand how PBD works. The reader should note that the performance pyramid is broken up into items required by the code and then items that are not included within the code. The Federal Emergency Management Agency (FEMA) describes the PBD process as an iterative process that begins with the selection of performance objectives, followed by the development of a preliminary design, an assessment as to whether or not the design meets the performance objectives, and finally redesign and reassessment, if required,

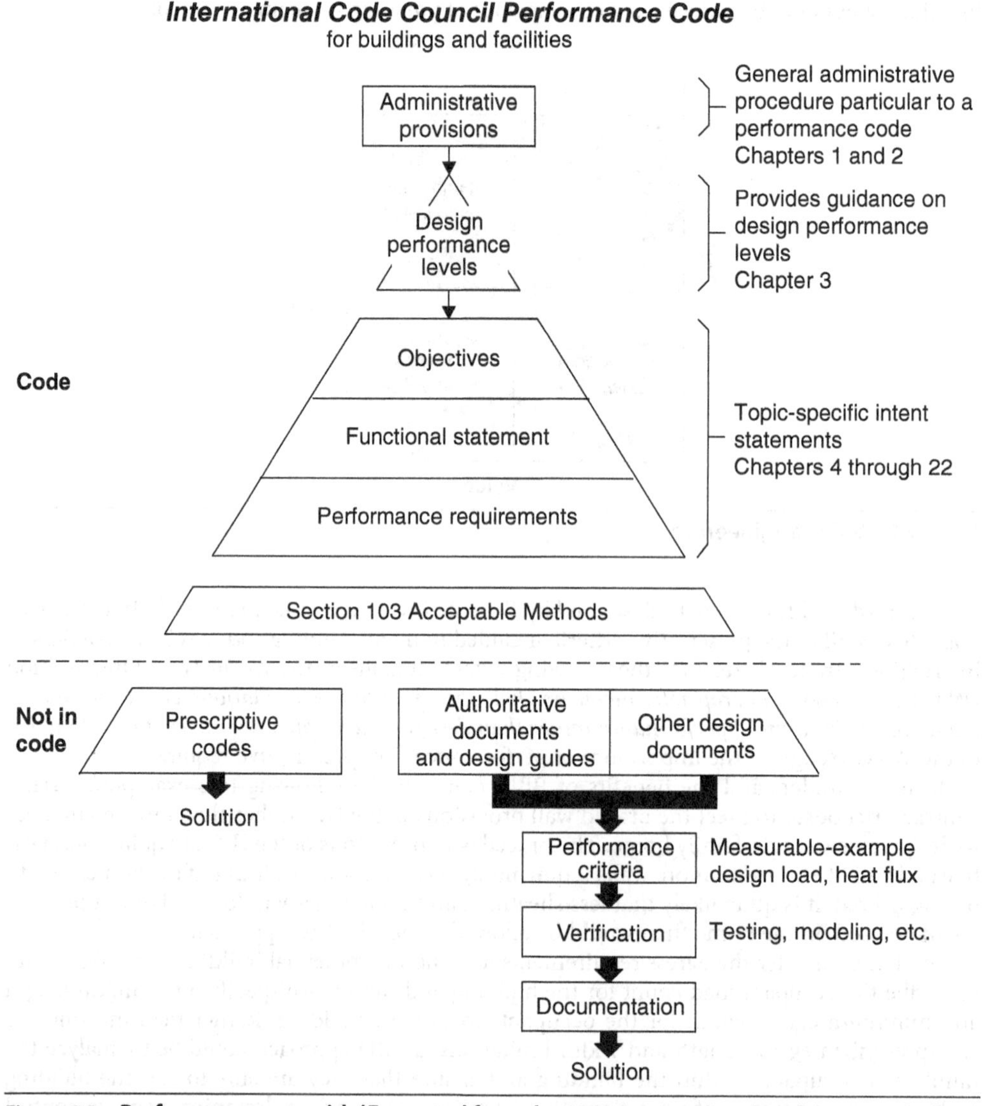

Figure 7.3 Performance pyramid. (*Excerpted from the 2018 International Performance Code; copyright 2017. Washington, D.C.: International Code Council.*)

until the desired performance level is achieved (FEMA 445, 2006). This process described by FEMA is displayed in the performance pyramid. Chapter 2 of FEMA P-424, "Design Guide for Improving School Safety in Earthquakes, Floods and High Winds," is an excellent resource for describing the PBD process.

7.3 *Scope*

Section 1301.1 of the IEBC identifies that the performance compliance method can only be used for alterations, additions, and changes in occupancy. Repairs are still required to comply with Chapter 4, while relocated buildings must comply with Chapter 14 of the IEBC. As has been discussed previously, the designer cannot follow the performance compliance method for some aspects of the design and either the prescriptive or work area method for others. Once starting down this path, the entire design must comply with Chapter 13 of the IEBC. Section 1301.1 states that the performance method is "*...intended to maintain or increase the current degree of public safety, health and general welfare in existing buildings.*" It allows some give and take in relation to some of the design disciplines as long as the overall safety score meets or exceeds the minimum requirement.

7.4 *Applicability*

In accordance with Section 1301.2, the performance compliance method can only be used for Groups A, B, E, F, I-2, M, R, and S occupancies. This method is not acceptable for Groups H, I-1, I-3, and I-4. While it is not clear why a hospital (Group I-2) would be allowed and a group home (Group I-1) is not, it makes sense why Group H and several I-occupancies are not allowed to use the performance compliance method. Establishing performance criteria for these hazardous and institutional occupancies can be difficult, and this criterion forms the basis of PBD. While the State of Wisconsin has adopted the IEBC for use, it has revised this section to clarify that the performance compliance method cannot be used for all H- and I-occupancies, thus removing Group I-2 as well (Wisconsin State Legislature, 2018).

Not only is the performance compliance method limited based upon use group but there are also certain restrictions based on the type of work being performed. The following subsections describe the limitations of the performance compliance method in relation to changes in occupancy, additions, and alterations.

7.4.1 Change in occupancy. When a portion of the building undergoes a change of occupancy, the entire building must be shown to comply with Chapter 13. As an alternative, only the changed portion is required to comply with this chapter, but it must be separated from the remaining occupancies by means of fire barriers or horizontal assemblies having a fire-resistance rating equal to that specified in Table 508.4 of the IBC (see Fig. 7.4).

As an example, consider a three-story office building that is fully fire sprinklered. The owner is proposing on changing the first-floor occupancy to a restaurant. Per Chapter 3 of the IBC, a restaurant is a Group A-2 occupancy, while the two upper floors of office space are considered Group B. Figure 7.4 shows that a 2-h separation is required between a restaurant and an office if the building is not sprinklered, and only a 1-h separation is required if it is sprinklered. Because this building is sprinklered, if the designer intends to only perform an analysis of the first floor, they would simply need to ensure that a 1-h horizontal separation is provided between the restaurant and second-floor office space.

REQUIRED SEPARATION OF OCCUPANCIES (HOURS)[f]																				
	A, E		I-1[a], I-3, I-4		I-2		R[a]		F-2, S-2[b], U		B[e], F-1, M, S-1		H-1		H-2		H-3, H-4		H-5	
OCCUPANCY	S	NS	S	NS	S	NS	S	NS	S	NS	S	NS	S	NS	S	NS	S	NS	S	NS
A, E	N	N	1	2	2	NP	1	2	N	1	1	2	NP	NP	3	4	2	3	2	NP
I-1[a], I-3, I-4	—	—	N	N	2	NP	1	NP	1	2	1	2	NP	NP	3	NP	2	NP	2	NP
I-2	—	—	—	—	N	N	2	NP	2	NP	2	NP	NP	NP	3	NP	2	NP	2	NP
R[a]	—	—	—	—	—	—	N	N	1[c]	2[c]	1	2	NP	NP	3	NP	2	NP	2	NP
F-2, S-2[b], U	—	—	—	—	—	—	—	—	N	N	1	2	NP	NP	3	4	2	3	2	NP
B[e], F-1, M, S-1	—	—	—	—	—	—	—	—	—	—	N	N	NP	NP	2	3	1	2	1	NP
H-1	—	—	—	—	—	—	—	—	—	—	—	—	N	NP	NP	NP	NP	NP	NP	NP
H-2	—	—	—	—	—	—	—	—	—	—	—	—	—	—	N	NP	1	NP	1	NP
H-3, H-4	—	—	—	—	—	—	—	—	—	—	—	—	—	—	—	—	1[d]	NP	1	NP
H-5	—	—	—	—	—	—	—	—	—	—	—	—	—	—	—	—	—	—	N	NP

S = Buildings equipped throughout with an automatic sprinkler system installed in accordance with Section 903.3.1.1.
NS = Buildings not equipped throughout with an automatic sprinkler system installed in accordance with Section 903.3.1.1.
N = No separation requirement.
NP = Not Permitted.

a. See Section 420.
b. The required separation from areas used only for private or pleasure vehicles shall be reduced by 1 hour but not to less than 1 hour.
c. See Section 406.3.2.
d. Separation is not required between occupancies of the same classification.
e. See Section 422.2 for ambulatory care facilities.
f. Occupancy separations that serve to define fire area limits established in Chapter 9 for requiring fire protection systems shall also comply with Section 707.3.10 and Table 707.3.10 in accordance with Section 901.7.

Figure 7.4 IBC Table 508.4. (*Excerpted from the 2018 International Building Code; copyright 2017. Washington, D.C.: International Code Council.*)

7.4.2 Additions. Section 1301.2.3 of the IEBC states that additions themselves must comply with the requirements of the *International Building Code*® (IBC®) or the *International Residential Code*® (IRC®). In addition, it requires that the combined height and area of the addition plus the existing structure meet the limitations prescribed in Chapter 5 of the IBC. The only exception to meeting the combined height and area restrictions would be to construct a fire wall conforming to Section 706 of the IBC, thus making the addition a separate building.

While the addition must meet current code requirements, in many cases it also causes significant alterations to be performed within the existing structure. Such alterations could include extending existing corridors, stairways that previously terminated outside now end within the building, and much more. It is often difficult to construct an addition without altering the existing structure.

7.4.3 Alterations. Alterations performed to existing buildings can in no way cause the building, or portion thereof, to become less safe or sanitary than the building is currently. The only exception to this rule is that Chapter 13 allows the level of safety or sanitation to be decreased if it can be shown that the altered state meets the requirements of the IBC. This will not likely occur in most existing buildings, as following the performance compliance path will most likely require multiple modifications to occur that will improve the level of safety and sanitation within the building rather than reducing it.

7.5 *Other Design Disciplines?*

The evaluation process included in Chapter 13 of the IEBC focuses on fire safety and means of egress. It also requires that a separate structural analysis be performed showing that the building complies with the structural requirements of the IBC. Fire safety, means of egress, and structural stability are important parameters, but they do not include all design disciplines that combine to make a building safe and sanitary.

The ICCPC provides a much more comprehensive approach to PBD. It outlines several design performance levels that are not addressed in the IEBC and includes performance requirements for the following design disciplines:

- Structural stability
- Fire safety
- Pedestrian circulation
- Safety of users
- Moisture
- Interior environment
- Mechanical
- Plumbing
- Fuel gas
- Electricity
- Energy efficiency
- Fire prevention
- Fire impact management
- Management of people

- Means of egress
- Emergency notification, access, and facilities
- Emergency responder safety
- Hazardous materials

While following the requirements of Chapter 13 in the IEBC is allowed, the designer may want to consider following the more holistic approach laid out in the ICCPC. The code official can allow this as an alternate means and methods as laid out in Section 104.11 of the IEBC, but the designer should confirm that this will be allowed by the code official prior to proceeding.

7.6 *Acceptance*

In order for the code official to accept a PBD as compliant, they will require an investigation and evaluation as described in the remainder of this chapter. In addition, the code official must ensure that all unsafe conditions are resolved, that the building complies with both the *International Fire Code*® (IFC®) and the *International Property Maintenance Code*® (IPMC®), and that it complies with the flood hazard provisions that have been discussed throughout this book.

It is difficult for one individual to ensure that all these items are appropriately addressed. The IFC is typically overseen by the local fire marshal rather than the building official. For existing buildings, the fire marshal will typically rely on Chapter 11 of the IFC. When submitting an application for a building permit, the fire marshal will typically perform a concurrent plan review of the construction documents along with the building department.

Oversight of the IPMC is typically provided by the code enforcement officer and not the building official. Section 101.2 of the IPMC describes its scope as applying to all existing structures "*...for light, ventilation, space, heating, sanitation, protection from elements, a reasonable level of safety from fire and other hazards, and for reasonable level of sanitary maintenance....*" The intent of the IPMC is to ensure the public health, safety, and welfare.

7.7 *Investigation and Evaluation*

Section 1301.4 requires that the building owner have the existing building investigated and evaluated. The remaining portion of Chapter 13 of the IEBC covers how this investigation and evaluation are to be performed. In the end, an evaluation report is to be submitted to the code official that identifies deficiencies and lists how compliance will be provided. Upon review of the report and any associated construction documents, the code official will determine if the proposed work complies with the IEBC, and if so, will issue a building permit subject to any reviews performed by other departments (see Sec. 7.6).

7.8 *Structural Analysis*

The performance compliance path outlined in the IEBC does not prescribe any PBD methodology for the structural analysis of the existing building. Rather, it specifically requires the analysis to show that the structural design complies with the requirements outlined in Chapter 16 of the IBC. In essence if the designer chooses to use the performance compliance method, they will need to show that the structural portion of the design meets the requirements for new construction.

Because of the language provided in this section, FEMA interprets the IEBC performance method as limited to relatively new buildings or buildings in which the structural system already complies with IBC requirements for new construction (FEMA, April 2017). This same view is held by the Existing Buildings Committee for the National Council of Structural Engineers Associations (NCSEA). This committee has stated that the performance compliance method is limited to newer buildings, whereas buildings that use obsolete structural systems would need to use either the prescriptive or work area compliance method (Bonowitz, 2017).

In the future Chapter 13 will likely incorporate performance methodologies for the structural analysis of existing buildings rather than simply referring to the IBC. Currently there are several PBD guidelines that are commonly used by structural engineers. Most of these guidelines are for unusually tall buildings or uncommon structural systems that are to be constructed in high-seismic regions. The most common guideline used for these types of applications is known as the TBI Guidelines for Performance-Based Seismic Design of Tall Buildings (PEER, 2017). This guideline was developed by the Pacific Earthquake Engineering Center (PEER) as an alternative to the prescriptive procedures outlined in the IBC and in ASCE 7.

Figure 7.5 displays an example of when the TBI Guidelines are commonly used. This image is of the 99 West Condominium Tower in downtown Salt Lake City, Utah. The building is roughly 375 ft in height and as of today is the third tallest building in the State of Utah. The structural system used consists of special reinforced concrete shear walls. When following the prescriptive procedures of the IBC and ASCE 7, a building with this type of structural system could not be taller than 160 ft. To show that the design met the intent of the code, the owner chose to follow the PBD procedures outlined in the TBI Guidelines. It should be noted that prior to proceeding down this path, the owner and SER should ensure that the code official would accept such an approach. Similar types of buildings exceeding the prescriptive height requirements of the code have been constructed throughout the United States following the TBI Guidelines.

In addition to the TBI Guidelines, PBD methodologies have been recommended by FEMA. In Chapter 2 of FEMA 424, there are several performance levels that are outlined for flood design, high-wind design, and seismic design (FEMA, December 2010). For seismic design FEMA 424 refers to ASCE 41, which addresses the seismic rehabilitation of existing buildings. ASCE 41 is also referenced in Chapter 3 of the IEBC as being applicable to all compliance methods. Unfortunately, as currently written in Section 303.3.1 of the IEBC, the PBD analysis of existing buildings would need to comply with the Tier 3 procedures outlined in ASCE 41 in addition to the specific performance objectives noted in Table 303.3.1.

For more than a decade, FEMA has provided funding to the Applied Technology Council (ATC) to develop new guidelines for performance-based structural design that could be used for both new and existing buildings. This work is being accomplished under ATC project number ATC-58. The purpose of the project is to advance the state of the practice in performance-based seismic design. The cost of the project in 2004 dollars was estimated to be around $21 million (FEMA 445, 2006). FEMA's goal is that the PBD procedures being developed will be incorporated into the National Earthquake Hazards Reduction Program's (NEHRP) Recommended Provisions for Seismic Regulations for New Buildings and Other Structures (Rojahn, 2014). These NEHRP regulations form the basis of the seismic provisions provided in the IBC. As such, the ultimate goal is that these PBD procedures will become a part of the adopted building code.

The PBD concepts that are provided in ASCE 41 are considered first-generation procedures, while the ATC-58 project is focused on developing the next generation of performance-based seismic design procedures. The methodology that is being developed is intended to be general enough that it can be applied to any building type, regardless of age, construction, or occupancy (FEMA P-58-2, 2012).

Figure 7.5 99 West Tower, Salt Lake City, Utah.

As part of this program, ATC has developed an electronic performance assessment calculation tool (PACT) that can be used to perform many of the probabilistic calculations and loss estimations that are described in the ATC-58 documents. Figure 7.6 displays a screenshot of the program. In reviewing the tabs at the top of the image, the reader can get a feel for the input that will be required as well as the output that can be provided. The risks associated with these

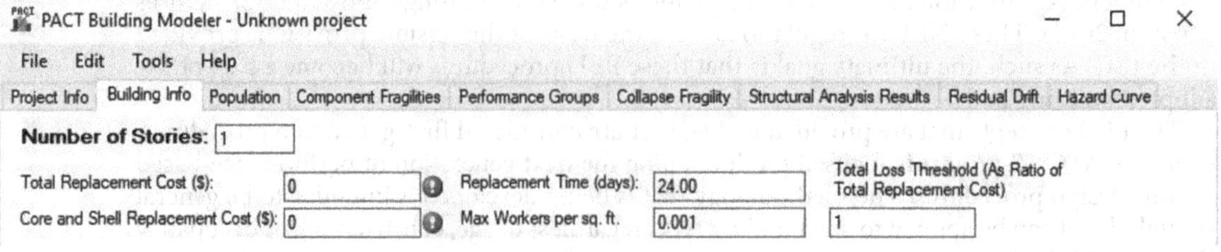

PACT Building Modeler - Unknown project

File Edit Tools Help

Project Info | Building Info | Population | Component Fragilities | Performance Groups | Collapse Fragility | Structural Analysis Results | Residual Drift | Hazard Curve

Number of Stories: 1

Total Replacement Cost ($): 0
Replacement Time (days): 24.00
Total Loss Threshold (As Ratio of Total Replacement Cost)
Core and Shell Replacement Cost ($): 0
Max Workers per sq. ft. 0.001
1

Figure 7.6 Performance assessment calculation tool (PACT).

PBD procedures can be defined in terms of potential for casualties, repair costs, and occupancy interruption (Rojahn, 2014). The PACT can be used to determine what those risks would be for the new or existing building being evaluated.

Figure 7.7 displays the performance assessment process that is being recommended as part of the ATC-58 project (FEMA P-58-2, 2012). There still appears to be no timeline for when this project will be completed or when it will be incorporated into the NEHRP guidelines. With that said, it is important to note that the design and regulating community see the need for PBD methods that can be readily used for the structural design of both new and existing buildings. For more information regarding the status of the ATC-58, or to download the PACT program, the reader should visit the ATC's website (https://atcouncil.org/).

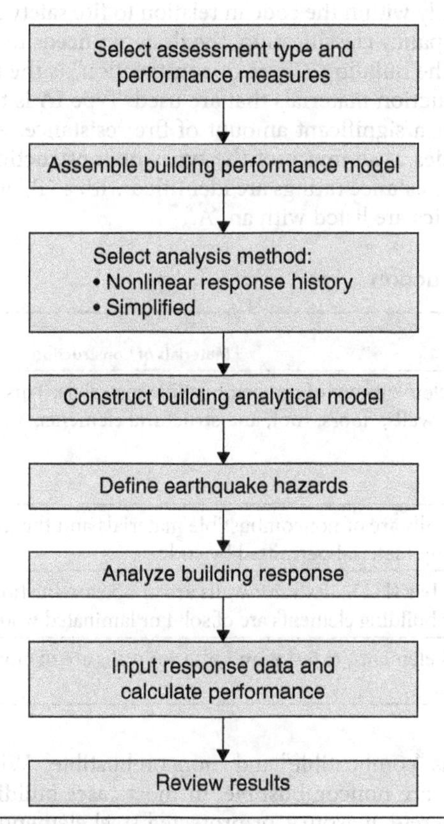

Figure 7.7 ATC-58, performance assessment process.

Future editions of the IEBC will likely incorporate improved methodologies for accomplishing the structural analysis under the performance compliance approach. In the meantime, it is up to the code official to determine if they would allow the structural analysis to be performed using something other than Chapter 16 of the IBC as an alternate means and method as outlined in Section 104.11 of the IEBC.

7.9 *Evaluation Process*

The evaluation requirements outlined in Chapter 13 deal with fire safety, means of egress, and general safety items. The following outlines in detail the items that are to be evaluated and what should be included in the evaluation report submitted to the code official.

7.9.1 Height and area. Chapter 5 of the IBC outlines the maximum height, number of floors, and floor area that a building can have. The size a building can be is contingent upon how the space will be used and the materials employed in its construction. As the size of the building increases, the construction materials used become more restrictive.

Chapter 3 of the IBC outlines the typical uses of a building. How the space will be utilized is described by the occupancy classification that is assigned. The occupancy classifications described in Chapter 3 of the IBC are shown in Table 3.1 of this text. It is important to note that each occupancy group is identified differently within the code in relation to fire safety and relative hazard.

In addition to the occupancy classification, the designer needs to determine what materials will be used to construct the building. "Type of construction" is the term used throughout the code to classify the construction materials that are used. Type IA is the most stringent form of construction providing for a significant amount of fire resistance, while Type VB is the least stringent. Table 7.1 provides a summary of the types of construction prescribed by the IBC. Those with standard fire-resistance ratings are identified with a "B," while those with additional fire-resistance characteristics are listed with an "A."

Table 7.1 Types of Construction

Type of Construction		Materials of Construction
I	IA IB	Building elements are of noncombustible materials. This includes exterior and interior walls, floors, roof, and structural elements.
II	IIA IIB	
III	IIIA IIIB	Exterior walls are of noncombustible materials and the interior building elements are of any material permitted by code.
IV	HT	Heavy timber (HT)—Exterior walls are of noncombustible materials and the interior building elements are of solid or laminated wood without concealed spaces.
V	VA VB	Structural elements, exterior and interior walls are of any materials permitted by code.

Table 7.1 uses the terms "combustible" and "noncombustible." While the IBC does not specifically list materials that are noncombustible, in most cases buildings that are noncombustible are composed of concrete, masonry, or protected steel elements. Should the code official need to determine if a material is noncombustible, they can request noncombustibility tests as described in Section 703.5 of the IBC.

While Table 7.1 provides a good description of each type of construction, perhaps the best way to differentiate between each type is to review Table 601 of the IBC. This table is shown in Fig. 7.8. Based upon the type of construction selected, the table specifies the fire-resistance rating of specific building elements. These building elements include the primary structural frame (i.e., columns, beams, girders, trusses, and spandrels), bearing walls (i.e., walls that have imposed loads on them), nonbearing walls, floors, and roofs. When using Table 601 the user must remember to review the footnotes as well. In many cases the fire-resistance rating can be decreased for certain elements.

FIRE-RESISTANCE RATING REQUIREMENTS FOR BUILDING ELEMENTS (HOURS)									
	TYPE I		TYPE II		TYPE III		TYPE IV	TYPE V	
BUILDING ELEMENT	A	B	A	B	A	B	HT	A	B
Primary structural frame[f] (see Section 202)	3[a,b]	2[a,b]	1[b]	0	1[b]	0	HT	1[b]	0
Bearing walls Exterior[e, f] Interior	 3 3[a]	 2 2[a]	 1 1	 0 0	 2 1	 2 0	 2 1/HT	 1 1	 0 0
Nonbearing walls and partitions Exterior	See Table 602								
Nonbearing walls and partitions Interior[d]	0	0	0	0	0	0	See Section 2304.11.2	0	0
Floor construction and associated secondary members (see Section 202)	2	2	1	0	1	0	HT	1	0
Roof construction and associated secondary members (see Section 202)	1½[b]	1[b,c]	1[b,c]	0[c]	1[b,c]	0	HT	1[b,c]	0

For SI: 1 foot = 304.8 mm.

a. Roof supports: Fire-resistance ratings of primary structural frame and bearing walls are permitted to be reduced by 1 hour where supporting a roof only.

b. Except in Group F-1, H, M and S-1 occupancies, fire protection of structural members in roof construction shall not be required, including protection of primary structural frame members, roof framing and decking where every part of the roof construction is 20 feet or more above any floor immediately below. Fire-retardant-treated wood members shall be allowed to be used for such unprotected members.

c. In all occupancies, heavy timber complying with Section 2304.11 shall be allowed where a 1-hour or less fire-resistance rating is required.

d. Not less than the fire-resistance rating required by other sections of this code.

e. Not less than the fire-resistance rating based on fire separation distance (see Table 602).

f. Not less than the fire-resistance rating as referenced in Section 704.10.

Figure 7.8 Table 601 of the IBC. (*Excerpted from the 2018 International Building Code; copyright 2017. Washington, D.C.: International Code Council.*)

The occupancy classification and type of construction establish the foundation upon which the majority of the building code is based. Chapter 3 of the IBC Commentary describes this as the theory of "equivalent risk." The concept is that an acceptable level of risk against the damages of fire can be obtained by limiting the height and area of buildings. The Commentary outlines the following three interdependent items in relation to equivalent risk:

1. The potential fire hazard associated with each occupancy classification
2. The reduction of fire hazard by limiting the floor areas and the height
3. The level of overall fire resistance provided by the type of construction

The greater the fire resistance rating noted in Table 601 of the IBC, the greater the allowable floor area and height of the building. The greater the potential fire hazards indicated by the occupancy classification, the lesser the allowable height and area of the facility. This concept of equivalent risk was developed by the insurance industry and has remained constant in the building codes for some time.

The performance compliance method provided in the IEBC incorporates the allowable numbers of stories, height, and area limitations from Chapter 5 of the IBC. To determine the building height score that will be used in the evaluation report, the design professional must first determine the score for the building height and then the score for the number of stories. The lesser of the two is the score that will be used in the evaluation report. Before the designer can perform the calculations, they will need to determine the allowable height and number of stories for the building based upon Chapter 5 of the IBC.

7.9.2 Compartmentation. Section 1301.6.3 of the IEBC requires that the design professional consider how the building has been divided up into separate fire areas by means of fire barriers or horizontal assemblies. Fire barriers and horizontal assemblies that are required as part of dwelling unit separations or corridor fire-resistance ratings are not to be included in this analysis. To be considered, the fire barriers and horizontal assemblies should be constructed as outlined in the IBC and should not have a fire-resistance rating of less than 2 h.

The design professional should use Table 1301.6.3 of the IEBC (see Fig. 7.9) to determine a compartmentation value (CV) for the building. This value should then be included in the "Building Code Summary Sheet" shown in Table 1301.7 of the IEBC. The value should be repeated in the fire safety, means of egress, and general safety categories in the table. An example will be provided at the end of this chapter.

COMPARTMENTATION VALUES					
	CATEGORIES				
OCCUPANCY	a Compartment size equal to or greater than 15,000 square feet	b Compartment size of 10,000 square feet	c Compartment size of 7,500 square feet	d Compartment size of 5,000 square feet	e Compartment size of 2,500 square feet or less
A-1, A-3	0	6	10	14	18
A-2	0	4	10	14	18
A-4, B, E, S-2	0	5	10	15	20
F, M, R, S-1	0	4	10	16	22
For SI: 1 square foot = 0.0929 m².					

Figure 7.9 Table 1301.6.3 of the IEBC. (*Excerpted from the 2018 International Existing Building Code; copyright 2017. Washington, D.C.: International Code Council.*)

7.9.3 Unit separations. Section 1301.6.4 of the IEBC requires that the design professional evaluate separations between tenant spaces, between dwelling units, or between sleeping rooms. The separation values shown in Table 1301.6.4 of the IEBC (see Fig. 7.10) should then be listed under the "Tenant and Dwelling Unit Separations" portion of the "Building Code Summary Sheet" (Table 1301.7 of the IEBC). The same value should be listed under each of the fire safety, means of egress, and general safety columns.

SEPARATION VALUES					
	CATEGORIES				
OCCUPANCY	a	b	c	d	e
A-1	0	0	0	0	1
A-2	−5	−3	0	1	3
R	−4	−2	0	2	4
A-3, A-4, B, E, F, M, S-1	−4	−3	0	2	4
I-2	0	1	2	3	4
S-2	−5	−2	0	2	4

Figure 7.10 Table 1301.6.4 of the IEBC. (*Excerpted from the 2018 International Existing Building Code; copyright 2017. Washington, D.C.: International Code Council.*)

The separation values that are provided in Table 1301.6.4 are divided into five separate categories—a through e. The following is a brief description of what each of these categories means and will help in selecting the appropriate values:

- *Category a:* This represents a condition where no fire separation exists, or incomplete separations are provided, or doors are not self-closing.
- *Category b:* Fire partitions and horizontal assemblies having a minimum 1-h fire-resistance rating and constructed in accordance with the IBC, or fire separations that do not comply with the IBC.
- *Category c:* Fire partitions having a minimum of 1-h fire-resistance rating and horizontal assemblies having a minimum of a 1-h fire-resistance rating. Each meets the construction requirements of the IBC.
- *Category d:* Fire barriers having a minimum of 1-h fire-resistance rating and horizontal assemblies having a minimum of a 2-h fire-resistance rating. Each meets the construction requirements of the IBC.
- *Category e:* Fire barriers or horizontal assemblies having a minimum of a 2-h fire-resistance rating and constructed in compliance with the IBC.

7.9.4 Corridor walls. Section 1301.6.5 of the IEBC requires the design professional to evaluate corridor walls for the degree of fire-resistance rating as well as completeness. The evaluation should compare the existing construction to what is required in Section 1020 of the IBC for new construction. The corridor wall values shown in Table 1301.6.5 of the IEBC (see Fig. 7.11) should then be listed under the "corridor walls" portion of the "Building Code Summary Sheet" (Table 1301.7 of the IEBC). The same value should be listed under each of the fire safety, means of egress, and general safety columns.

CORRIDOR WALL VALUES				
	CATEGORIES			
OCCUPANCY	a	b	c[a]	d[a]
A-1	−10	−4	0	2
A-2	−30	−12	0	2
A-3, F, M, R, S-1	−7	−3	0	2
A-4, B, E, S-2	−5	−2	0	5
I-2	−10	0	1	2
a. Corridors not providing at least one-half the exit access travel distance for all occupants on a floor shall use Category b.				

Figure 7.11 Table 1301.6.5 of the IEBC. ***(Excerpted from the 2018 International Existing Building Code; copyright 2017. Washington, D.C.: International Code Council.)***

The corridor wall values listed in Table 1301.6.5 of the IEBC are divided into four separate categories—a through d. The following is a brief description of what each of these categories means and will help the design professional in selecting the appropriate values for the evaluation:

- *Category a:* This represents a condition where no fire-resistance rating is provided, fire partitions are incomplete, or doors opening into the corridor are not self-closing.
- *Category b:* Walls have less than 1-h fire-resistance rating or they have not been constructed in accordance with the IBC.
- *Category c:* Fire partitions have a minimum of 1-h fire-resistance rating, but less than 2-h, and are constructed in accordance with the IBC.
- *Category d:* Fire partitions have a 2-h fire-resistance rating and are constructed in accordance with the IBC, including openings into corridors.

7.9.5 Vertical openings. This is one of the trickier parts of the performance evaluation that is performed by the design professional. Section 1301.6.6 of the IEBC requires the design professional to perform an evaluation of interior exit stairways or ramps, hoistways, escalator openings, and other shaft enclosures. There are actually three parts to this evaluation before a value can be input into the "Building Code Summary Sheet."

The first step is to select the protection value of these openings based upon Table 1301.6.6(1) of the IEBC (see Fig. 7.12). The second step is to select the construction-type factor from Table 1301.6.6(2) of the IEBC (see Fig. 7.12). The final step is to multiply the two factors together and to enter the sum into the "Building Code Summary Sheet." If it is a one-story building, or if unenclosed vertical openings conform to Section 713 of the IBC, a value of +2 should be entered. In no case should the listed value be greater than +2.

7.9.6 Heating, ventilation, and air conditioning (HVAC) systems. Section 1301.6.7 of the IEBC requires the design professional to evaluate the ability of the HVAC system to resist the movement of smoke and fire beyond the point of origin. Five separate categories are provided when evaluating the HVAC system as described next:

- *Category a (Points = −10):* This represents a condition where mechanical plenums are not provided in accordance with Section 602 of the *International Mechanical Code®* (IMC®).
- *Category b (Points = −5):* This category exists when the air movement provided at egress elements does not conform to Section 1020.5 of the IBC.
- *Category c (Points = −15):* This category occurs when both Categories a and b exist.

VERTICAL OPENING PROTECTION VALUE	
PROTECTION	**VALUE**
None (unprotected opening)	−2 times number of floors connected
Less than 1 hour	−1 times number of floors connected
1 to less than 2 hours	1
2 hours or more	2

(a)

CONSTRUCTION-TYPE FACTOR									
FACTOR	TYPE OF CONSTRUCTION								
	IA	IB	IIA	IIB	IIIA	IIIB	IV	VA	VB
	1.2	1.5	2.2	3.5	2.5	3.5	2.3	3.3	7

(b)

Figure 7.12 **(*a*) Table 1301.6.6(1) and (*b*) Table 1301.6.6(2) of the IEBC. (*Excerpted from the 2018 International Existing Building Code; copyright 2017. Washington, D.C.: International Code Council.*)**

- *Category d (Points = 0):* This category applies to HVAC systems that comply with both Section 1020.5 of the IBC and Section 602 of the IMC.
- *Category e (Points = +5):* This condition only applies to mechanical systems that are only serving one story, or for buildings utilizing a boiler/chiller system that connects two or more stories.

If the facility being evaluated is a Group I-2 occupancy and Categories a, b, or c would apply, the evaluation automatically fails. In other words, Group I-2 occupancies can only follow the performance compliance path if the HVAC system conforms to Category d or e.

7.9.7 Automatic fire detection. As described in Section 1301.6.8 of the IEBC, the design professional must also provide a detailed review of the existing fire detection systems as part of the evaluation. These systems should be checked for their location and operation as required by the IBC and IMC for new construction. The fire detection values shown in Table 1301.6.8 of the IEBC (see Fig. 7.13) should then be listed under the "automatic fire detection" portion of the

AUTOMATIC FIRE DETECTION VALUES						
OCCUPANCY	CATEGORIES					
	a	b	c	d	e	f
A-1, A-3, F, M, R, S-1	−10	−5	0	2	6	NA
A-2	−25	−5	0	5	9	NA
A-4, B, E, S-2	−4	−2	0	4	8	NA
I-2	NP	NP	NP	4	5	2

NA = Not Applicable.
NP = Not Permitted.

Figure 7.13 **Table 1301.6.8 of the IEBC. (*Excerpted from the 2018 International Existing Building Code; copyright 2017. Washington, D.C.: International Code Council.*)**

"Building Code Summary Sheet" (Table 1301.7 of the IEBC). The same value should be listed under each of the fire safety, means of egress, and general safety columns.

The fire detection values listed in Table 1301.6.8 are divided into six separate categories—a through f. The following is a brief description of what each of these categories means and will help the design professional in selecting the appropriate values for the evaluation. Just as with HVAC systems, if the facility being evaluated is a Group I-2 occupancy and Categories a, b, or c would apply, the evaluation automatically fails.

- *Category a:* No fire detection is provided.
- *Category b:* Smoke detectors are provided within the HVAC system and are maintained in accordance with the IFC.
- *Category c:* Smoke detectors are provided within the HVAC system and are installed in compliance with the requirements for new buildings as specified in the IMC.
- *Category d:* Smoke detectors are provided throughout all floor areas other than individual dwelling and sleeping units or tenant spaces.
- *Category e:* Smoke detectors are installed throughout the floor area.
- *Category f:* Smoke detectors are installed within the corridor only.

7.9.8 Fire alarm systems. Now it is time for the design professional to evaluate the fire alarm system that is, or will be, provided. The evaluation should compare how these systems comply with the requirements in Section 907 of the IBC, and the appropriate fire alarm values from Table 1301.6.9 of the IEBC (see Fig. 7.14) should be selected. This value should be listed under the "fire alarm system" portion of the "Building Code Summary Sheet" (Table 1301.7 of the IEBC). The following describes the four separate categories that are included in Table 1301.6.9:

- *Category a:* A fire alarm system is not provided.
- *Category b:* A fire alarm system that includes manual fire alarm boxes and notification appliances complying with Sections 907.4 and 908.5.2 of the IBC.
- *Category c:* A complete fire alarm system complying with Section 907 of the IBC.
- *Category d:* This includes a Category c fire alarm system as well as an emergency voice/alarm communication system and a fire command center as outlined in the IBC.

FIRE ALARM SYSTEM VALUES				
	CATEGORIES			
OCCUPANCY	a	b[a]	c	d
A-1, A-2, A-3, A-4, B, E, R	−10	−5	0	5
F, M, S	0	5	10	15
I-2	−4	1	2	5

a. For buildings equipped throughout with an automatic sprinkler system, add 2 points for activation by a sprinkler water-flow device.

Figure 7.14 Table 1301.6.9 of the IEBC. (*Excerpted from the 2018 International Existing Building Code; copyright 2017. Washington, D.C.: International Code Council.*)

7.9.9 Smoke control. Section 1301.6.10 of the IEBC requires the design professional to evaluate the ability of smoke control systems to control the movement of smoke. These systems can be either passive or active (e.g., mechanical systems). For smoke control systems, there are

SMOKE CONTROL VALUES						
	CATEGORIES					
OCCUPANCY	a	b	c	d	e	f
A-1, A-2, A-3	0	1	2	3	6	6
A-4, E	0	0	0	1	3	5
B, M, R	0	2[a]	3[a]	3[a]	3[a]	4[a]
F, S	0	2[a]	2[a]	3[a]	3[a]	3[a]
I-2	−4	0	0	0	3	0
a. This value shall be 0 if compliance with Category d or e in Section 1301.6.8.1 has not been obtained.						

Figure 7.15 Table 1301.6.10 of the IEBC. ***(Excerpted from the 2018 International Existing Building Code; copyright 2017. Washington, D.C.: International Code Council.)***

six different categories that are used to determine the appropriate value from Table 1301.6.10 of the IEBC (see Fig. 7.15). The selected value should then be input under the "smoke control" portion of the "Building Code Summary Sheet" (Table 1301.7 of the IEBC). The following describes the six separate categories that are included in Table 1301.6.10:

- *Category a:* A smoke control system is not provided.
- *Category b:* The building is fully sprinklered and has evenly distributed operable openings equivalent to 20 ft^2 for every 50 linear feet of exterior wall.
- *Category c:* Building has openings as listed in Category b and has one enclosed stairway that is accessed by each floor. The stairway shall have operable exterior windows.
- *Category d:* Building has openings as listed in Category b and there is at least one smoke-proof enclosure.
- *Category e:* The building is fully sprinklered, and air-handling systems are designed to accomplish smoke containment. Return and exhaust air is moved directly to the outside, and a minimum of six air changes per hour are provided at each floor.
- *Category f:* Each stairway shall be either a smokeproof enclosure (per Section 1023.11 of the IBC), shall be pressurized (per Section 909.20.5 of the IBC), or shall have operable exterior windows.

7.9.10 Means of egress. One of the most important aspects of the performance evaluation is to analyze the means of egress system. Per Section 1301.6.11 of the IEBC, the design professional must evaluate the means of egress capacity and the number of exits that are available to the building occupants. This evaluation should compare what is available to the requirements of IBC Sections 1003.7, 1004, 1005, 1006, 1007, 1016.2, 1026.1, 1028.2, 1028.5, 1029.2, 1029.4, and 1030. Existing fire escapes can be accepted as a component of the means of egress system; however, new fire escapes serving as part of the means of egress cannot be constructed under the performance approach.

Table 1301.6.11 of the IEBC (see Fig. 7.16) provides the applicable values that should be input into the "Building Code Summary Sheet" (Table 1301.7 of the IEBC). There are five separate categories to consider when selecting the score. The following describes each of the five categories:

- *Category a:* Compliance with the required means of egress capacity, or with the required number of exits, is achieved with the use of an existing fire escape.

MEANS OF EGRESS VALUES[a]					
OCCUPANCY	CATEGORIES				
	a	b	c	d	e
A-1, A-2, A-3, A-4, E, I-2	−10	0	2	8	10
M	−3	0	1	2	4
B, F, S	−1	0	0	0	0
R	−3	0	0	0	0

a. The values indicated are for buildings six stories or less in height. For buildings over six stories above grade plane, add an additional −10 points.

Figure 7.16 Table 1301.6.11 of the IEBC. ***(Excerpted from the 2018 International Existing Building Code; copyright 2017. Washington, D.C.: International Code Council.)***

- *Category b:* The capacity of the means of egress complies with Section 1005 of the IBC, and the minimum number of exists complies with Section 1006 of the IBC.
- *Category c:* The capacity of the means of egress is equal to, or exceeds, 125 percent of the required capacity, the means of egress complies with the minimum required width of the IBC, and the number of exits conforms to Section 1006 of the IBC.
- *Category d:* The number of exits provided exceeds the minimum number required by Section 1006 of the IBC, and they are located a sufficient distance apart to comply with the provisions of Section 1007 of the IBC.
- *Category e:* The area being evaluated complies with both Categories c and d.

7.9.11 Dead ends. The design professional must review the building and locate spaces that are required to be served by more than one means of egress and to determine whether there are areas where the occupants are confined to a single path of travel. Based upon the length of confined travel, an appropriate value should be selected from Table 1301.6.12 of the IEBC (see Fig. 7.17) and input into the "Building Code Summary Sheet" (Table 1301.7 of the IEBC). There are four separate categories to evaluate these dead ends, each of which is described below.

- *Category a:* The dead end distance is equal to no more than 35 ft in nonsprinklered buildings and 70 ft in sprinklered buildings.

DEAD-END VALUES				
OCCUPANCY	CATEGORIES[a]			
	a	b	c	d
A-1, A-3, A-4, B, F, M, R, S	−2	0	2	−4
A-2, E	−2	0	2	−4
I-2	−2	0	2	−6

a. For dead-end distances between categories, the dead-end value shall be obtained by linear interpolation.

Figure 7.17 Table 1301.6.12 of the IEBC. ***(Excerpted from the 2018 International Existing Building Code; copyright 2017. Washington, D.C.: International Code Council.)***

- *Category b:* The dead end distance is equal to no more than 20 ft, although a distance of 50 ft is acceptable for Group B occupancies in accordance with Section 1020.4, Exception 2 of the IBC.
- *Category c:* No dead ends exist, or the ratio of the length to the width is less than 2.5:1.0.
- *Category d:* The dead ends exceed the maximum lengths of Category a.

7.9.12 Maximum exit travel distance. Section 1301.6.13 of the IEBC requires that the design professional evaluate travel distance to an approved exit throughout the space. This should be compared to the maximum allowable travel distance prescribed by Section 1017.1 of the IBC. To determine the value that should be input into the "Building Code Summary Sheet" (Table 1301.7 of the IEBC), the design professional should use Equation 13-6 of the IEBC, which is provided below.

$$\text{Points} = 20 \times \frac{\text{Max. allowable travel} - \text{Max. actual travel}}{\text{Max. allowable travel}}$$

7.9.13 Elevator control. Section 1301.6.14 requires the design professional to evaluate the elevator equipment and controls that are available to the fire department. In accordance with the IFC, elevators should be equipped with emergency recall and in-car operation controls. The values to be input into the "Building Code Summary Sheet" should be obtained from Table 1301.6.14 of the IEBC (see Fig. 7.18). A value of "0" should be input if the building is a single story. The following provides a description of the four categories listed in Table 1301.6.14:

- *Category a:* No elevator is provided.
- *Category b:* Any elevator does not have Phase I emergency recall or Phase II emergency in-car operation.
- *Category c:* All elevators have Phase I emergency recall and Phase II emergency in-car operation as required by the IFC.
- *Category d:* All elevators meet Category c and at least one elevator complies with the new construction requirements and serves all floors.

ELEVATOR CONTROL VALUES				
	CATEGORIES			
ELEVATOR TRAVEL	a	b	c	d
Less than 25 feet of travel above or below the primary level of elevator access for emergency fire-fighting or rescue personnel	−2	0	0	+2
Travel of 25 feet or more above or below the primary level of elevator access for emergency fire-fighting or rescue personnel	−4	NP	0	+4

For SI: 1 foot = 304.8 mm.
NP = Not Permitted.

Figure 7.18 Table 1301.6.14 of the IEBC. ***(Excerpted from the 2018 International Existing Building Code; copyright 2017. Washington, D.C.: International Code Council.)***

7.9.14 Means of egress emergency lighting. The design professional should not only evaluate the presence of means of egress lighting but also the reliability of such lighting in accordance with Section 1301.6.15 of the IEBC. Based upon this determination, an appropriate value should be selected from Table 1301.6.15 of the IEBC (see Fig. 7.19) and input into the "Building Code Summary Sheet." The following describes the three categories associated with means of egress emergency lighting as listed in Table 1301.6.15:

- *Category a:* Means of egress lighting and exit signs *are not* provided with emergency power as required by Section 2702 of the IBC.
- *Category b:* Means of egress lighting and exit signs *are* provided with emergency power as required by Section 2702 of the IBC.
- *Category c:* Emergency power is provided to means of egress lighting and exit signs.

MEANS OF EGRESS EMERGENCY LIGHTING VALUES			
NUMBER OF EXITS REQUIRED BY SECTION 1006 OF THE INTERNATIONAL BUILDING CODE	CATEGORIES		
	a	b	c
Two or more exits	NP	0	4
Minimum of one exit	0	1	1

NP = Not Permitted.

Figure 7.19 Table 1301.6.15 of the IEBC. (*Excerpted from the 2018 International Existing Building Code; copyright 2017. Washington, D.C.: International Code Council.*)

7.9.15 Mixed occupancies. When the building has two or more occupancies, the design professional must evaluate the separation provided between the occupancies. An appropriate value must be selected from Table 1301.6.16 of the IEBC (see Fig. 7.20) and then input into the "Building Code Summary Sheet." If no separation exists, a value of "0" should be input. The following describes each of the three categories identified in Table 1301.6.16:

- *Category a:* Occupancies are separated by a minimum 1-h fire barrier, 1-h horizontal assembly, or both.
- *Category b:* Separations between occupancies conform to Section 508.4 of the IBC.
- *Category c:* Separations between occupancies have a fire-resistance rating of at least twice that required by Section 508.4 of the IBC.

MIXED OCCUPANCY VALUES[a]			
OCCUPANCY	CATEGORIES		
	a	b	c
A-1, A-2, R	−10	0	10
A-3, A-4, B, E, F, M, S	−5	0	5
I-2	NP	0	5

NP = Not Permitted.
a. For fire-resistance ratings between categories, the value shall be obtained by linear interpolation.

Figure 7.20 Table 1301.6.16 of the IEBC. (*Excerpted from the 2018 International Existing Building Code; copyright 2017. Washington, D.C.: International Code Council.*)

7.9.16 Automatic sprinklers. Section 1301.6.17 of the IEBC requires that existing sprinkler systems be evaluated and that an appropriate score be selected from Table 1301.6.17 of the IEBC (see Fig. 7.21). The selected value should then be input under the "automatic sprinklers" portion of the "Building Code Summary Sheet" (Table 1301.7 of the IEBC). The "means of egress" value should be divided by 2 prior to entering it into Table 1301.7.

SPRINKLER SYSTEM VALUES						
	CATEGORIES					
OCCUPANCY	a[a]	b[a]	c	d	e	f
A-1, A-3, F, M, R, S-1	−6	−3	0	2	4	6
A-2	−4	−2	0	1	2	4
A-4, B, E, S-2	−12	−6	0	3	6	12
I-2	NP	NP	NP	8	10	NP

NP = Not Permitted.
a. These options cannot be taken if Category a in Section 1301.6.18 is used.

Figure 7.21 **Table 1301.6.17 of the IEBC. (*Excerpted from the 2018 International Existing Building Code; copyright 2017. Washington, D.C.: International Code Council.*)**

If a high-rise building is undergoing change of occupancy to Group R, it shall be provided with sprinklers throughout in accordance with Sections 403 and 903 of the IBC. Group I-2 occupancies that are categorized as a, b, c, or f shall be considered as failing the evaluation. The following describes the six separate categories that are included in Table 1301.6.17:

- *Category a:* Sprinklers are required throughout; however, sprinkler protection is not provided, or the system provided is not adequate for the hazard to be protected.
- *Category b:* Sprinklers are required for a portion of the building; however, sprinkler protection is not provided, or the system provided is not adequate for the hazard to be protected.
- *Category c:* Sprinklers are not required and none are provided.
- *Category d:* Sprinklers are required for a portion of the building and are provided in that portion only. The system complied with the code at the time of original construction and has been maintained and supervised per the IBC.
- *Category e:* Sprinklers are required throughout, and they have been provided throughout. The system complies with Chapter 9 of the IBC.
- *Category f:* Sprinklers are not required throughout, but they have been provided throughout in accordance with Chapter 9 of the IBC.

7.9.17 Standpipes. The design professional should evaluate whether the building would require standpipes in accordance with Section 905 of the IBC and should compare what is provided to what is required. An appropriate value should be selected from Table 1301.6.18 of the IEBC (see Fig. 7.22) and input into the "Building Code Summary Sheet." The following describes the four separate categories listed in Table 1301.6.18:

- *Category a:* Standpipes are required but no standpipes are provided, or the standpipes provided do not comply with Section 905.3 of the IBC.

STANDPIPE SYSTEM VALUES				
OCCUPANCY	CATEGORIES			
	a[a]	b	c	d
A-1, A-3, F, M, R, S-1	−6	0	4	6
A-2	−4	0	2	4
A-4, B, E, S-2	−12	0	6	12
I-2	−2	0	1	2
a. This option cannot be taken if Category a or Category b in Section 1301.6.17 is used.				

Figure 7.22 Table 1301.6.18 of the IEBC. (*Excerpted from the 2018 International Existing Building Code; copyright 2017. Washington, D.C.: International Code Council.*)

- *Category b:* Standpipes are not required and none are provided.
- *Category c:* Standpipes *are* required and are provided in accordance with Section 905 of the IBC.
- *Category d:* Standpipes *are not* required but they are provided in accordance with Section 905 of the IBC.

7.9.18 Incidental uses. Section 1301.6.19 of the IEBC requires that the design professional evaluate any incidental use areas within the building. These areas should be compared to what is required for new construction in Section 509.4.2 of the IBC, and the appropriate values should be selected from Table 1301.6.19 of the IEBC (see Fig. 7.23) and input into the "Building Code Summary Sheet."

INCIDENTAL USE AREA VALUES							
PROTECTION REQUIRED BY TABLE 509 OF THE *INTERNATIONAL BUILDING CODE*	PROTECTION PROVIDED						
	None	1 hour	AS	AS with CRS	1 hour and AS	2 hours	2 hours and AS
2 hours and AS	−4	−3	−2	−2	−1	−2	0
2 hours, or 1 hour and AS	−3	−2	−1	−1	0	0	0
1 hour and AS	−3	−2	−1	−1	0	−1	0
1 hour	−1	0	−1	−1	0	0	0
1 hour, or AS with CRS	−1	0	−1	−1	0	0	0
AS with CRS	−1	−1	−1	−1	0	−1	0
1 hour or AS	−1	0	0	0	0	0	0
AS = Automatic Sprinkler System; CRS = Construction capable of resisting the passage of smoke (see IBC Section 509.4.2 of the *International Building Code*).							

Figure 7.23 Table 1301.6.19 of the IEBC. (*Excerpted from the 2018 International Existing Building Code; copyright 2017. Washington, D.C.: International Code Council.*)

The lowest score among all the incidental use areas should be input into the "Building Code Summary Sheet." If the building does not include an incidental use area, a value of "0" should be input. Values for incidental use areas should also not be input if the building is required to be sprinklered throughout. In reviewing Fig. 7.23, "AS" signifies that automatic sprinklers are provided, and "CRS" signifies that the construction is capable of resisting the passage of smoke.

7.9.19 Smoke compartmentation. Smoke compartmentation is only required for I-2 occupancies. In accordance with Section 1301.6.20 of the IEBC, if I-2 occupancies are not broken up into separate smoke compartments that are equal to or less than 22,500 square feet, they are considered to fail the evaluation. If compartmentation is provided that meets this limitation, a value of zero should be placed in the "smoke compartmentation" row under fire safety, means of egress, and general safety.

7.9.20 Patient ability, concentration, smoke compartment location, and ratio to attendant. This section also applies only to Group I-2 occupancies. The design professional is required to evaluate each smoke compartment separately for the ability of the patients for self-preservation, for the concentration of patients, and for the attendant-to-patient ratio. The scores for each of these three items must be multiplied together and if the sum is greater than 9, the evaluation has failed. The following is a breakdown of how each of these three scores is determined:

- *Patient Ability for Self-Preservation:* This score is obtained from Table 1301.6.21.1 of the IEBC (see Fig. 7.24). Category a is used for patients who are mobile and considered to be capable of self-preservation. Category b is for patients who rely on assistance for evacuation, and Category c is for patients who cannot be evacuated.

PATIENT ABILITY VALUES			
OCCUPANCY	CATEGORIES		
	a	b	c
I-2	1	2	3

Figure 7.24 Table 1301.6.21.1 of the IEBC. (*Excerpted from the 2018 International Existing Building Code; copyright 2017. Washington, D.C.: International Code Council.*)

- *Patient Concentration:* This score is obtained from Table 1301.6.21.2 of the IEBC (see Fig. 7.25). Category a is smoke compartments having 10 or fewer patients. Category b is for smoke compartments having 10 to 40 patients, and Category c is for compartments having more than 40 patients.

PATIENT CONCENTRATION VALUES			
OCCUPANCY	CATEGORIES		
	a	b	c
I-2	1	2	3

Figure 7.25 Table 1301.6.21.2 of the IEBC. (*Excerpted from the 2018 International Existing Building Code; copyright 2017. Washington, D.C.: International Code Council.*)

- *Attendant-to-Patient Ratio:* This score is obtained from Table 1301.6.21.3 of the IEBC (see Fig. 7.26). Category a is used for attendant-to-patient ratios of 1:5 or less. Category b is for a ratio of 1:6 to 1:10, and Category c is for a ratio of greater than 1:10.

ATTENDANT-TO-PATIENT RATIO VALUES			
OCCUPANCY	CATEGORIES		
	a	b	c
I-2	1	2	3

Figure 7.26 **Table 1301.6.21.3 of the IEBC. (*Excerpted from the 2018 International Existing Building Code; copyright 2017. Washington, D.C.: International Code Council.*)**

7.10 *Building Score*

The design professional should now sum all the values included in the "Building Code Summary Sheet" to come up with the calculated fire safety (FS), calculated means of egress (ME), and calculated general safety (GS). The calculated values for each must be greater than, or equal to, the mandatory fire safety (MFS), mandatory means of egress (MME), and mandatory general safety (MGS) values that are provided in Table 1301.8 of the IEBC (see Fig. 7.27).

MANDATORY SAFETY SCORES[a]			
OCCUPANCY	FIRE SAFETY (MFS)	MEANS OF EGRESS (MME)	GENERAL SAFETY (MGS)
A-1	20	31	31
A-2	21	32	32
A-3	22	33	33
A-4, E	29	40	40
B	30	40	40
F	24	34	34
I-2	19	34	34
M	23	40	40
R	21	38	38
S-1	19	29	29
S-2	29	39	39

a. MFS = Mandatory Fire Safety.
MME = Mandatory Means of Egress.
MGS = Mandatory General Safety.

Figure 7.27 **Table 1301.8 of the IEBC. (*Excerpted from the 2018 International Existing Building Code; copyright 2017. Washington, D.C.: International Code Council.*)**

Figure 7.28 provides an example of how the "Building Code Summary Sheet" (e.g., Table 1301.7 of the IEBC) is filled out. The example is for a Group B occupancy, and per Table 1301.8 the mandatory values should be equal to 30, 40, and 40, respectively. This example is for a building constructed rather recently and whose allowable height and area are significantly less than what

TABLE 1301.7
SUMMARY SHEET—BUILDING CODE

Existing occupancy: Sample Project
Proposed occupancy: Office - Group B
Year building was constructed: 2005
Number of stories: 2 Height in feet: 22' - 6"
Type of construction: Type V-B
Area per floor: 6,900 square feet
Percentage of open perimeter increase: 100 %
Completely suppressed: Yes X No ___
Corridor wall rating: 1-hour
Type: Fire partitions
Compartmentation: Yes ___ No X
Required door closers: Yes X No ___
Fire-resistance rating of vertical opening enclosures: ___
Type of HVAC system: Complies with 2018 IMC, serving number of floors: 2
Automatic fire detection: Yes X No ___
Type and location: Water flow and smoke detection
Fire alarm system: Yes X No ___
Type: Manual
Smoke control: Yes ___ No X
Type: NA
Adequate exit routes: Yes X No ___
Dead ends: Yes ___ No X
Maximum exit access travel distance: ___
Elevator controls: Yes X No ___
Means of egress emergency lighting: Yes X No ___
Mixed occupancies: Yes ___ No X
Standpipes: Yes ___ No X
Patient ability for self-preservation: NA
Incidental use: Yes ___ No X
Patient concentration: NA
Smoke compartmentation less than 22,500 sq. feet (2092 m²): Yes ___ No X
Attendant-to-patient ratio: NA

SAFETY PARAMETERS	FIRE SAFETY (FS)	MEANS OF EGRESS (ME)	GENERAL SAFETY (GS)
1301.6.1 Building height	10	10	10
1301.6.2 Building area	9	9	9
1301.6.3 Compartmentation	5	5	5
1301.6.4 Tenant and dwelling unit separations	0	0	0
1301.6.5 Corridor walls	0	0	0
1301.6.6 Vertical openings	1	1	1
1301.6.7 HVAC systems	0	0	0
1301.6.8 Automatic fire detection	0	0	0
1301.6.9 Fire alarm system	-3	-3	-3
1301.6.10 Smoke control	* * * *	0	0
1301.6.11 Means of egress	* * * *	0	0
1301.6.12 Dead ends	* * * *	2	2
1301.6.13 Maximum exit access travel distance	* * * *	6	6
1301.6.14 Elevator control	0	0	0
1301.6.15 Means of egress emergency lighting	* * * *	0	0
1301.6.16 Mixed occupancies	0	* * * *	0
1301.6.17 Automatic sprinklers	12	÷ 2 = 6	12
1301.6.18 Standpipes	0	0	0
1301.6.19 Incidental use	0	0	0
1301.6.20 Smoke compartmentation	0	0	0
1301.6.21.1 Patient ability for self-preservation[a]	* * * *	0	0
1301.6.21.2 Patient concentration[a]	* * * *	0	0
1301.6.21.3 Attendant-to-patient ratio[a]	* * * *	0	0
Building score—total value	34	36	42
* * * *No applicable value to be inserted. a. Only applicable to Group I-2 occupancies.	>30, Okay	**<40, No Good!**	>40, Okay

Figure 7.28 Sample Building Code Summary Sheet.

is allowed by the current code. Even though the construction is relatively new, the PBD analysis shows that this building would not pass, as the calculated values at the bottom of the summary sheet do not exceed the mandatory values in each case. In addition to getting the fire safety, means of egress, and general safety items to work out, the design professional will still need to show that the structure fully conforms to the current IBC.

CHAPTER 8

RELOCATED OR MOVED BUILDINGS

8.1 *Introduction*

The previous versions of the *International Existing Building Code*® (IEBC®) had different requirements for moved structures depending on whether the designer was following the prescriptive, work area, or performance compliance method. The prescriptive provisions required that the moved structure comply with the *International Building Code*® (IBC®) in its entirety. The performance compliance method required the structural systems to comply with the IBC but allowed the performance method to be used for fire safety, means of egress, and general safety requirements. The work area method required the building to be in compliance with the *International Fire Code*® (IFC®) and *International Property Maintenance Code*® (IPMC®); the foundation to be designed per the IBC; and the structural design to meet the wind, seismic, snow, and flood loads for the new location (with a few exceptions).

In the 2018 IEBC, the relocated and moved building requirements have been removed from the individual compliance methods and are now compiled into a single chapter. Chapter 14 must now be used in every case that an existing building will be moved or relocated. The chapter itself is almost identical to the provisions that were previously included under the work area compliance method. With that said, it is important to understand that this is a stand-alone chapter that must be complied with regardless of the compliance path that is chosen by the design professional. In many cases, a construction project involving a moved building will also include repairs, alterations, or changes in use. If such is the case, the designer can choose from the following two options:

1. They can choose to design the entire moved and altered building to comply with the provisions of the IBC.
2. They can meet the requirements of Chapter 14 in the IEBC for the relocated structure, comply with Chapter 4 for any needed repairs, and then choose one of the three compliance methods discussed previously in this book to address any alterations or additions being made.

While Chapter 14 of the IEBC addresses both relocated and moved buildings, it is important to understand the difference between the two. Figure 8.1 shows a building that was constructed at one site and is now being relocated to a new site, likely to make room for a new development. This is a moved structure. Moved structures are typically site-built, have permanent foundations, and are likely intended to remain on that site for the life of the building.

While moving buildings does happen from time to time, the code official will more often be dealing with relocated buildings. Section 202 of the IBC defines relocatable buildings as "*A partially or completely assembled building constructed and designed to be reused multiple times and transported to different building sites.*" Relocatable buildings are often considered modular, or prefabricated, structures and are constructed off-site under controlled plant conditions. They are constructed in such a manner that makes the transportation, or relocation, of the structure much easier. Often these buildings will spend years in one location and then be moved to several other locations throughout their lifetime.

Relocatable buildings can include construction trailers, modular classrooms, mobile homes, tiny homes, and much more (see Fig. 8.2). There is sometimes debate as to whether these types of buildings should be held to the strict provisions of the building code. With the tiny home revolution, this debate has come to the forefront. To address this topic, the National Fire Protection Association® (NFPA®) Building Code Development Committee has performed quite a bit of research on the topic. In a white paper developed by the committee entitled "Applying Building Codes to Tiny Homes" they identified the following two conditions that could allow a

Figure 8.1 Relocation of a historic church in Salem, Massachusetts. (*Wikimedia, 2009.*)

Figure 8.2 Modular classroom. (*Anchor, n.d.*)

relocatable building to not fall under the purview of the building codes, and therefore the IEBC (BCDC, 2017):

- *Recreational Vehicles:* If the unit can be placed on a permanent trailer chassis with wheels, it is possible to argue that it does not fall within the scope of the building codes (see Fig. 8.3). In this instance, the unit would need to remain in a mobile-ready state and would likely need to comply with other laws or regulations that are established by the state's Department of Motor Vehicles. Some design standards such as NFPA 1192, Standard on Recreational Vehicles, are often referenced for the design of recreational vehicles.

Figure 8.3 Tiny home on wheels. (*Drucker, 2015.*)

- *Manufactured Homes:* Manufactured homes are regulated by the U.S. Department of Housing and Urban Development (HUD) under the "Mobile Home and Construction Safety Standards Act of 1974." Most states are considered local regulators and may enforce additional items above and beyond the HUD standards such as adopting NFPA 501, Standard for Manufactured Housing. Manufactured homes must have a label from HUD affixed to it identifying that it meets HUD regulations. Such structures are not held to the provisions of the building codes but also should not be altered or relocated without the approval of a local HUD regulator. Figure 8.4 shows a carport that was added to an existing mobile home. While doing work of this nature might be common, no such work should occur without written approval from HUD.

Figure 8.4 Carport addition to mobile home. (*Superior Awning, n.d.*)

Owners, design professionals, and code officials should know that the requirements for manufactured housing vary from state to state. While the NFPA Building Code Development Committee does include manufactured housing as something that may fall out of the purview of the building code, some states specifically adopt the International codes for manufactured housing. As an example, the Texas Industrialized Housing and Buildings has specifically adopted the International code family for buildings, modules, and modular components.

While there is a significant difference between moved and relocatable structures, the provisions in Chapter 14 apply to both equally. Many assume that relocatable buildings should not be held to the same standard, but unless the code official feels that it falls under the purview of a recreational vehicle or a manufactured home, the IEBC requirements should be met. The following sections will cover in detail what the IEBC requirements are for relocated or moved buildings.

8.2 *Scope and Conformance*

As noted in the title of this chapter, Chapter 14 of the IEBC provides the requirements for relocated or moved structures. Section 1401.1 of the IEBC also clarifies that the chapter applies to "relocatable buildings" discussed above. In order to conform to the IEBC, relocated or moved structures must comply with the following:

1. The provisions of the IFC and IPMC.
2. Repairs, alterations, and changes of occupancy must comply with the IEBC.
3. New construction must comply with the IBC or *International Residential Code*® (IRC®).
4. Must comply with the provisions of Section 1402 of the IEBC.

At this point in the book, items 2 and 3 above should be clear to the reader. Item 4 will be addressed in Secs. 8.3 through 8.7 of this chapter.

8.3 *Location on the Lot*

Section 1402.1 of the IEBC requires that the moved or relocated building be placed on the site in a manner that complies with either the IBC or the IRC. The rest of this chapter will compare the provisions in both the IBC and IRC as they apply to Chapter 14 of the IEBC. When doing so it is important that the reader understand that the IBC can be used for all buildings, while the IRC applies only to one- and two-family dwellings and townhomes that are no more than three stories in height.

8.3.1 Site plan. Section 107 of the IBC and Section R106 of the IRC describe what the permit applicant must include within the construction documents that are submitted for a building permit. Both the IBC and IRC require that a detailed site plan be included within the construction documents. They both require that the following information be shown on the site plan:

- Size and location of relocated or moved building
- Location of existing structures on lot
- Distances from lot lines
- Design flood elevations (flood hazard areas only)
- Any construction to be demolished

In addition to what is noted above, the IBC requires that the existing street grade be shown, that the proposed finished grade be specified, and that it be drawn in accordance with an accurate boundary line survey. Figure 8.5 provides a sample site plan that would meet the minimum

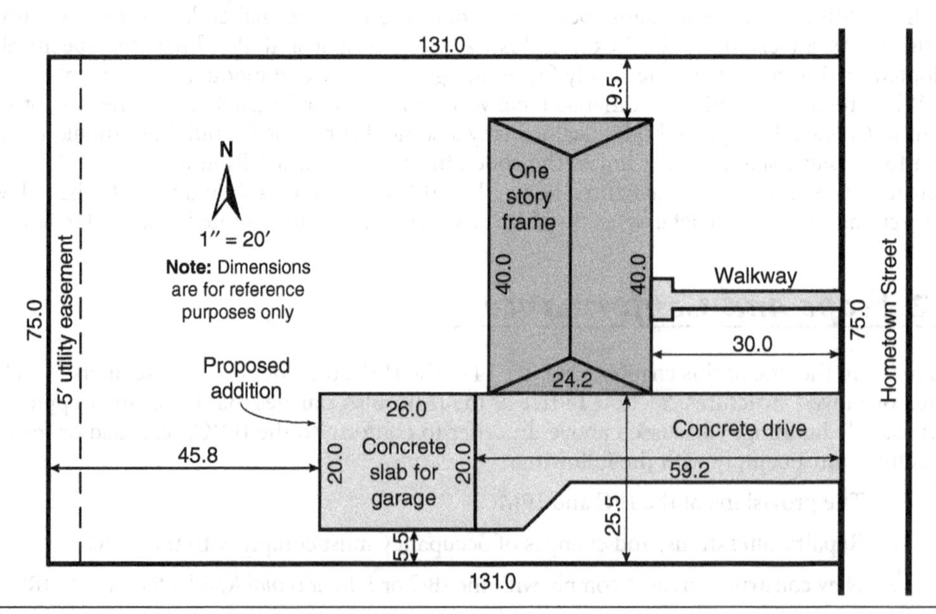

Figure 8.5 Sample site plan.

provisions of the IRC. This site plan is used in several building guides that have been developed by the Colorado Chapter of the International Code Council® (ICC®) (CCICC, 2016).

Many local jurisdictions may require additional information to be provided on the site plan as well. While the site plan is important to show compliance with many of the building code provisions, tighter regulations are often enforced by the local planning or zoning department. As an example, the Whatcom County Planning and Development Services Department in the State of Washington requires that the following additional items also be shown on the site plan (Whatcom, n.d.):

- Location and dimensions of all easements
- Location and dimensions of driveway encroachments
- Direction of water flow
- Critical areas (i.e., streams, creeks, and wetlands)
- Vicinity map showing route to site
- Existing and proposed wells, sewage systems, utility lines, and fuel tanks
- Erosion control measures

The location of the structure on the site can trigger other requirements in the building code, such as requiring fire-rated walls on the exterior or not allowing openings, as will be discussed below. The site plan serves as a tool for the code official to determine if the requirements of both the building code and local zoning ordinances will be met.

8.3.2 Proximity to lot line. One of the key elements of reviewing the site plan is to see how close the structure in question will be to the lot line, commonly referred to as the property line. If the structure is too close to the lot line, or to another building located on the same lot, it could require that elements of the relocated or moved building add fire-resistive elements. Table 602 of the IBC prescribes when the exterior wall is required to be fire rated (see Fig. 8.6).

There are three factors that must be considered when reviewing Table 602 of the IBC. First is the actual fire separation distance. Section 202 of the IBC defines the fire separation distance as ...*the distance from the building face to one of the following:*

- *The closest interior lot line*
- *The centerline of the street, alley, or public way*
- *An imaginary line between two buildings on the lot*

Once the fire separation distance has been determined, the second consideration is the type of construction for the building in question. The final factor is how the building will be used, or the occupancy group. After obtaining these three pieces of information, the designer is then able to determine if a fire rating is required for the exterior walls.

For an example consider a modular school building of Type VB construction. It will be placed 8 ft from the property line, and its use will be Group E. When reviewing Table 602 of the IBC (see Fig. 8.6) a 1-h fire rating is listed. As such, the designer would need to ensure that the existing exterior wall that is perpendicular to the property line already has a 1-h rating or will need to provide construction details for how the 1-h rating will be provided.

In addition to the fire rating, the IBC regulates the size and types of openings that are allowed in exterior walls. Table 705.8 of the IBC (see Fig. 8.7) provides limitations for openings in exterior walls that are within 30 ft of a property line. As with the fire-rating requirements provided in Table 602 of the IBC, if the exterior face of the building is 30 ft or greater from the property line, there are no limitations as to the amount or size of openings.

Like the exterior wall fire rating, there are also three factors that must be considered when determining the opening limitations from Table 705.8 of the IBC. The first is to determine the

IRE-RESISTANCE RATING REQUIREMENTS FOR EXTERIOR WALLS BASED ON FIRE SEPARATION DISTANCE[a, d, g]				
FIRE SEPARATION DISTANCE = X (feet)	TYPE OF CONSTRUCTION	OCCUPANCY GROUP H[e]	OCCUPANCY GROUP F-1, M, S-1[f]	OCCUPANCY GROUP A, B, E, F-2, I, R[i], S-2, U[h]
$X < 5^b$	All	3	2	1
$5 \le X < 10$	IA Others	3 2	2 1	1 1
$10 \le X < 30$	IA, IB IIB, VB Others	2 1 1	1 0 1	1^c 0 1^c
$X \ge 30$	All	0	0	0

For SI: 1 foot = 304.8 mm.
a. Load-bearing exterior walls shall also comply with the fire-resistance rating requirements of Table 601.
b. See Section 706.1.1 for party walls.
c. Open parking garages complying with Section 406 shall not be required to have a fire-resistance rating.
d. The fire-resistance rating of an exterior wall is determined based upon the fire separation distance of the exterior wall and the story in which the wall is located.
e. For special requirements for Group H occupancies, see Section 415.6.
f. For special requirements for Group S aircraft hangars, see Section 412.3.1.
g. Where Table 705.8 permits nonbearing exterior walls with unlimited area of unprotected openings, the required fire-resistance rating for the exterior walls is 0 hours.
h. For a building containing only a Group U occupancy private garage or carport, the exterior wall shall not be required to have a fire-resistance rating where the fire separation distance is 5 feet (1523 mm) or greater.
i. For a Group R-3 building of Type II-B or Type V-B construction, the exterior wall shall not be required to have a fire-resistance rating where the fire separation distance is 5 feet (1523 mm) or greater.

Figure 8.6 Table 602 of the IBC. (*Excerpted from the 2018 International Building Code; copyright 2017. Washington, D.C.: International Code Council.*)

fire separation distance. The next step is to note whether the building will be protected by fire sprinklers. The last item is to decide whether the openings in the exterior wall will be protected. Section 202 of the IBC defines opening protectives as either a fire door, fire shutter, or fire window assembly that is located within a fire-rated wall. Each of these obviously increases cost. Specific requirements for each are outlined in Chapter 7 of the IBC.

For the exterior wall opening example, consider the same modular classroom. In accordance with Section 903.2.3 of the IBC, it will not require fire sprinklers due to the small floor area and low number of students. It is likely that the existing openings do not consist of fire doors or fire windows, so the designer should first determine how much area of unprotected openings is allowed. By referring to Table 705.8 of the IBC (see Fig. 8.7), unprotected openings in an exterior wall of an unsprinklered building that is located 8 ft from the property line cannot exceed 10 percent of the wall area. If the existing openings in the exterior wall near the property line exceed 10 percent, the designer will need to either provide protected openings or move the building farther from the property line.

It is also important to ensure that exterior projections comply with Section 705.2 of the IBC. This section does not allow projections to extend beyond the exterior wall if the wall is within 2 ft of the property line. If the fire separation distance becomes 3 ft, a 12-in. projection is allowed. Once the fire separation distance becomes 5 ft, a 40-in. projection becomes acceptable.

All projections that are within 5 ft of the property line and consist of combustible materials are required to have a 1-h fire rating. In lieu of a 1-h rating heavy timber construction or fire-retardant-treated lumber can be used. Projections in buildings of Type I or II construction must consist of noncombustible materials. For relocated or moved buildings, the code official must ensure that projections that will be located within 5 ft of the property line meet these requirements.

MAXIMUM AREA OF EXTERIOR WALL OPENINGS BASED ON FIRE SEPARATION DISTANCE AND DEGREE OF OPENING PROTECTION		
FIRE SEPARATION DISTANCE (feet)	DEGREE OF OPENING PROTECTION	ALLOWABLE AREA[a]
0 to less than 3[b, c, k]	Unprotected, Nonsprinklered (UP, NS)	Not Permitted[k]
	Unprotected, Sprinklered (UP, S)[i]	Not Permitted[k]
	Protected (P)	Not Permitted[k]
3 to less than 5[d, e]	Unprotected, Nonsprinklered (UP, NS)	Not Permitted
	Unprotected, Sprinklered (UP, S)[i]	15%
	Protected (P)	15%
5 to less than 10[e, f, j]	Unprotected, Nonsprinklered (UP, NS)	10%[h]
	Unprotected, Sprinklered (UP, S)[i]	25%
	Protected (P)	25%
10 to less than 15[e, f, g, j]	Unprotected, Nonsprinklered (UP, NS)	15%[h]
	Unprotected, Sprinklered (UP, S)[i]	45%
	Protected (P)	45%
15 to less than 20[f, g, j]	Unprotected, Nonsprinklered (UP, NS)	25%
	Unprotected, Sprinklered (UP, S)[i]	75%
	Protected (P)	75%
20 to less than 25[f, g, j]	Unprotected, Nonsprinklered (UP, NS)	45%
	Unprotected, Sprinklered (UP, S)[i]	No Limit
	Protected (P)	No Limit
25 to less than 30[f, g, j]	Unprotected, Nonsprinklered (UP, NS)	70%
	Unprotected, Sprinklered (UP, S)[i]	No Limit
	Protected (P)	No Limit
30 or greater	Unprotected, Nonsprinklered (UP, NS)	No Limit
	Unprotected, Sprinklered (UP, S)[i]	No Limit
	Protected (P)	No Limit

For SI: 1 foot = 304.8 mm.
UP, NS = Unprotected openings in buildings not equipped throughout with an automatic sprinkler system in accordance with Section 903.3.1.1.
UP, S = Unprotected openings in buildings equipped throughout with an automatic sprinkler system in accordance with Section 903.3.1.1.
P = Openings protected with an opening protective assembly in accordance with Section 705.8.2.

a. Values indicated are the percentage of the area of the exterior wall, per story.
b. For the requirements for fire walls of buildings with differing heights, see Section 706.6.1.
c. For openings in a fire wall for buildings on the same lot, see Section 706.8.
d. The maximum percentage of unprotected and protected openings shall be 25 percent for Group R-3 occupancies.
e. Unprotected openings shall not be permitted for openings with a fire separation distance of less than 15 feet for Group H-2 and H-3 occupancies.
f. The area of unprotected and protected openings shall not be limited for Group R-3 occupancies, with a fire separation distance of 5 feet or greater.
g. The area of openings in an open parking structure with a fire separation distance of 10 feet or greater shall not be limited.
h. Includes buildings accessory to Group R-3.
i. Not applicable to Group H-1, H-2 and H-3 occupancies.
j. The area of openings in a building containing only a Group U occupancy private garage or carport with a fire separation distance of 5 feet or greater shall not be limited.
k. For openings between S-2 parking garage and Group R-2 building, see Section 705.3, Exception 2.

Figure 8.7 Table 705.8 of the IBC. (*Excerpted from the 2018 International Building Code; copyright 2017. Washington, D.C.: International Code Council.*)

The IRC also has provisions that sometimes require the exterior wall or projection to be fire-rated, limit the extent of projections, and limit the amount of exterior wall openings. Table R302.1(1) of the IRC (see Fig. 8.8) provides the requirements for these items in relation to buildings that are not sprinklered, while a similar table is provided for buildings that are sprinklered. As one would assume, the requirements for the sprinklered building are significantly less.

EXTERIOR WALLS			
EXTERIOR WALL ELEMENT		MINIMUM FIRE-RESISTANCE RATING	MINIMUM FIRE SEPARATION DISTANCE
Walls	Fire-resistance rated	1 hour—tested in accordance with ASTM E119, UL 263 or Section 703.3 of the *International Building Code* with exposure from both sides	0 feet
	Not fire-resistance rated	0 hours	≥ 5 feet
Projections	Not allowed	NA	< 2 feet
	Fire-resistance rated	1 hour on the underside, or heavy timber, or fire- retardant-treated wood[a, b]	≥ 2 feet to < 5 feet
	Not fire-resistance rated	0 hours	≥ 5 feet
Openings in walls	Not allowed	NA	< 3 feet
	25% maximum of wall area	0 hours	3 feet
	Unlimited	0 hours	5 feet
Penetrations	All	Comply with Section R302.4	< 3 feet
		None required	3 feet

For SI: 1 foot = 304.8 mm.
NA = Not Applicable.
a. The fire-resistance rating shall be permitted to be reduced to 0 hours on the underside of the eave overhang if fireblocking is provided from the wall top plate to the underside of the roof sheathing.
b. The fire-resistance rating shall be permitted to be reduced to 0 hours on the underside of the rake overhang where gable vent openings are not installed.

Figure 8.8 Table R302.1(1) of the IRC. (*Excerpted from the 2018 International Residential Code; copyright 2017. Washington, D.C.: International Code Council.*)

The main purpose of the exterior wall fire ratings, opening limitations, and projection limitations is to reduce the potential for fires within the building to spread to adjacent structures. Today's building codes incorporate multiple passive fire protection measures that have helped reduce many of the catastrophic fires that we have experienced in the past. Ensuring that the exterior walls or projections of a relocated or moved building meet these requirements is an essential task of the code official.

8.3.3 Foundations near slopes. It is important to ensure that foundations for relocated or moved buildings are not placed near slopes that exceed one unit vertical in three units horizontal (i.e., 33% slope). The provisions noted in Section R403.1.7 of the IRC and Section 1808.7 of the IBC are close to identical when dealing with slopes. There are three methods that both codes prescribe when placing footings near either ascending or descending slopes. The separation distance between the building and an ascending slope is referred to as the building "clearance," while the distance from a descending slope is termed the building "setback."

The first and most common method is shown in Fig. 8.9. This figure is taken from the IRC, but an identical one is provided in the IBC. This method can only be used when the slope does not exceed one unit vertical in one unit horizontal (i.e., 100 percent slope). If a relocated building will be placed above a slope, the exterior footing must be placed no closer to the edge of the slope than one-third the height of the slope, but in no case does that distance need to exceed 40 ft. If the building will be placed at the bottom of a slope, the footing must have a clearance of one half the height of the slope but is not required to exceed 15 ft.

For buildings next to slopes that exceed one unit vertical in one unit horizontal, they must follow an alternate approach from what is shown in Fig. 8.9. This approach has separate procedures for ascending and descending slopes. For ascending slopes, it requires that a horizontal line be drawn from the top of the foundation toward the slope. A line is then drawn at 45° that is tangent to the slope itself. A sufficient clearance is provided if the 45° line does not pass through the foundation. This approach is depicted in Fig. 8.10.

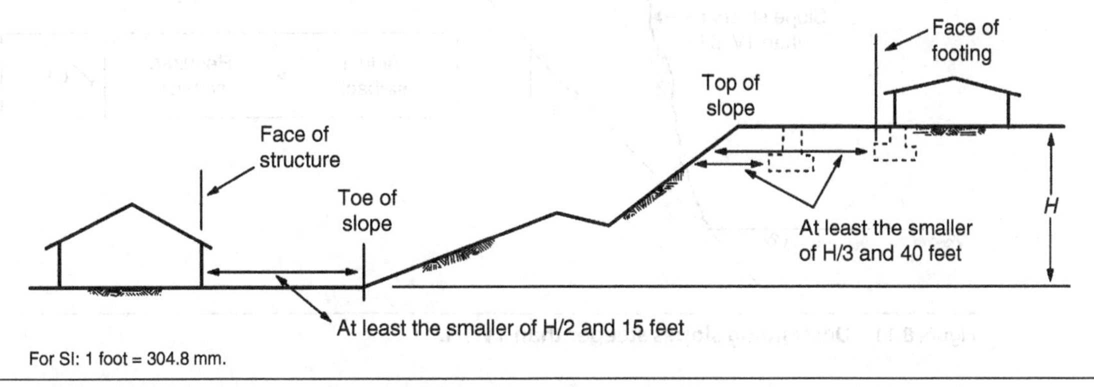

Figure 8.9 Figure R403.1.7.1 of the IRC. (*Excerpted from the 2018 International Residential Code; copyright 2017. Washington, D.C.: International Code Council.*)

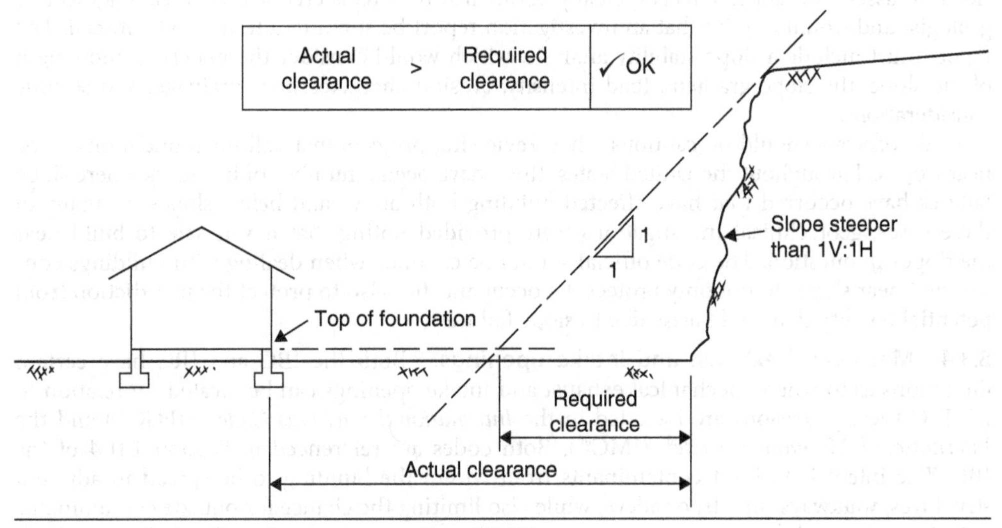

Figure 8.10 Ascending slopes steeper than 1V:1H.

For descending slopes, the grade plane above the slope serves as the horizontal line, and then a 45° line is drawn tangentially from the toe of the slope. The building must be set back a sufficient distance to ensure that the 45° line does not pass through the foundation as shown in Fig. 8.11.

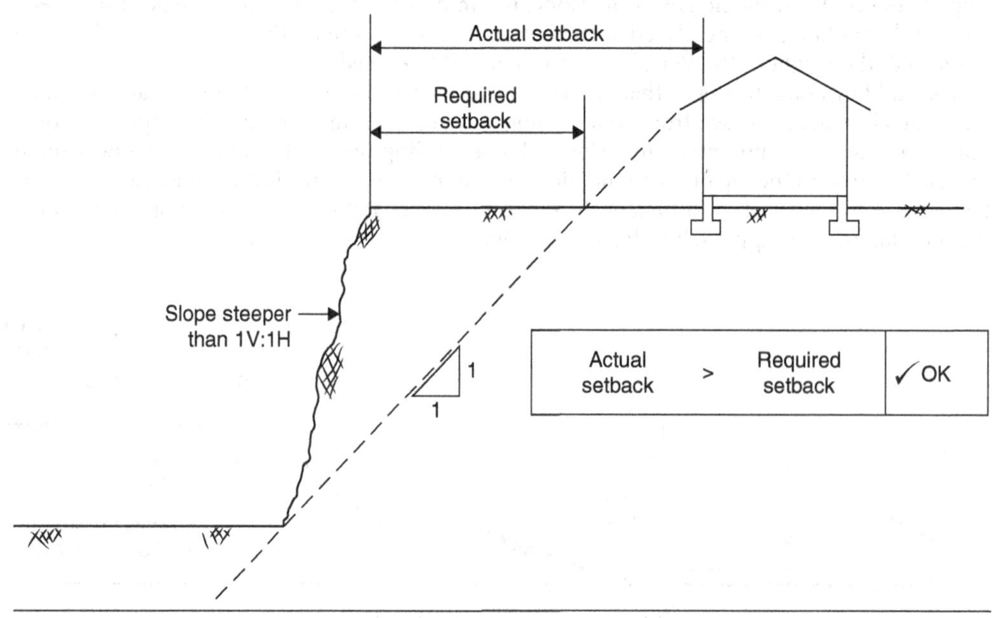

Figure 8.11 Descending slopes steeper than 1V:1H.

If the relocated building owner does not choose to follow the clearance or setback requirements outlined in the IRC or IBC for slopes, they can elect to hire a licensed design professional to assess the slope. This is typically performed by a registered geotechnical engineer or geologist and would require that an investigation report be submitted to the code official. The report must include a slope stability analysis, which would consider the materials and height of the slope, the slope gradient, load intensity, erosion characteristics, drainage, and seismic considerations.

Code officials should be cautious when reviewing projects that call for foundations to be near slopes. Throughout the United States, there have been a number of instances where slope failures have occurred that have affected building both above and below slopes. In many of these cases, geotechnical investigations were provided noting that it was safe to build near the slopes in question. The code official should be cautious when dealing with buildings constructed near slopes to not only protect the occupants but also to protect the jurisdiction from potential liability that could arise due to slope failures.

8.3.4 Mechanical exhaust and intake openings. Both the IBC and IRC have certain limitations as to where mechanical exhaust and intake openings can be located. In relation to the IBC these provisions are included in the *International Fuel Gas Code®* (IFGC®) and the *International Mechanical Code®* (IMC®). Both codes are referenced in Section 101.4 of the IBC. The intent is to limit contaminants from within the building to be spread to adjacent structures, walkways, streets, or alleys, while also limiting the chance for outside contaminants to be brought inside the building.

While not as important as the exterior wall fire-resistance requirements discussed previously, the location of intake and exhaust openings can have a significant impact on persons within and outside of relocated buildings. As such, the code official should ensure that such openings will be located so as to meet code requirements.

In summary, every permit application for a relocated or moved building must be accompanied by a site plan complying with Sec. 8.3.1 discussed above. Based upon the information on the site plan, the code official must determine whether the exterior walls are required to be fire rated and if the existing openings and projections associated with those walls meet with the applicable IBC and IRC limitations. All slopes should also be identified, and foundations must be placed so as to meet the setback and clearance requirements of the IBC and IRC. Finally, the proposed location of the building on the lot should not adversely affect the code limitations in relation to the locations of mechanical exhaust and intake openings.

8.4 *Foundation*

Section 1402.2 of the IEBC requires that the foundation supporting the relocated or moved building comply with the requirements of either the IBC or IRC. It also states that the connection of the building to the foundation is to be in accordance with these codes. The foundation and anchorage requirements in each code are quite different, so each will be addressed separately.

8.4.1 ***International Residential Code.*** Chapter 4 of the IRC outlines the footing and foundation requirements for one- and two-family buildings and townhomes that are not more than three stories. Before discussing the requirements of Chapter 4, it is important to understand the restrictions of the IRC.

The IRC provides a prescriptive approach to residential design. If the project meets the limitations of the code, the user simply needs to follow the recipe, and voila, a building can be constructed. So, what are those limitations? Without covering them all in detail, the following is a partial list. If one of these applies to the project, then the design should be performed in accordance with the IBC, one of the other design standards referenced in the IRC, or accepted engineering practice.

- High-wind regions along the East Coast as well as the coast of Alaska. Per Figure R301.2(4)B of the IRC, these are areas with a design wind speed of 140 mi/h or greater.
- Special wind regions that are either highlighted in Figure R301.2(4)B of the IRC or where the local jurisdiction requires a design wind speed that is greater than specified in the figure. It is assumed that this only applies where the design wind speed in these regions reaches 140 mi/h or greater.
- While the IRC applies to high-seismic regions, it cannot be used for structures located in Seismic Design Category E.
- In high-seismic regions the weights of materials are often limited. An example would be that wood-framed exterior walls cannot exceed 15 lb/ft^2.
- Structures with odd shapes or irregular configurations are limited in high-seismic regions. A listing of seven such configurations is defined in Section R301.2.2.2.5 of the IRC.
- The IRC considers a maximum ground snow load of 70 lb/ft^2. This is appropriate for most regions, but areas that receive excessive snow would not apply.

The user must be aware of the limitations noted above before taking the time to design the footings and foundations using Chapter 4 of the IRC. To size the footings, it is important to determine how much load the site soils can support. For this purpose, the IRC provides Table R401.4.1, which is shown in Fig. 8.12. This table lists common soil types in one column and then notes the allowable soil-bearing pressure in the next column. The allowable soil-bearing pressure is the key value in determining the size of the footings. Note that the lowest value in the table is 1500 lb/ft^2. Unless a geotechnical investigation is performed, most jurisdictions will require that the footings be sized using the 1500 lb/ft^2 value.

PRESUMPTIVE LOAD-BEARING VALUES OF FOUNDATION MATERIALS[a]	
CLASS OF MATERIAL	LOAD-BEARING PRESSURE (pounds per square foot)
Crystalline bedrock	12,000
Sedimentary and foliated rock	4,000
Sandy gravel and/or gravel (GW and GP)	3,000
Sand, silty sand, clayey sand, silty gravel and clayey gravel (SW, SP, SM, SC, GM and GC)	2,000
Clay, sandy, silty clay, clayey silt, silt and sandy siltclay (CL, ML, MH and CH)	1,500[b]

For SI: 1 pound per square foot = 0.0479 kPa.
a. Where soil tests are required by Section R401.4, the allowable bearing capacities of the soil shall be part of the recommendations.
b. Where the building official determines that in-place soils with an allowable bearing capacity of less than 1,500 psf are likely to be present at the site, the allowable bearing capacity shall be determined by a soils investigation.

Figure 8.12 Table R401.4.1 of the IRC. (*Excerpted from the 2018 International Residential Code; copyright 2017. Washington, D.C.: International Code Council.*)

As with all tables provided in the codes, it is important to read the footnotes. Footnote "b" in Table R401.4.1 notes that the code official may require a geotechnical investigation if they feel that the soils are not able to support 1500 lb/ft^2. Sections R401.2 and R401.4 clarify when a geotechnical investigation may be required by the code official. This includes anytime that the site soils might be expansive, compressible, shifting, or otherwise questionable. The IRC also notes that if fill soils will be placed below the footings and foundations, the installation and testing requirements should be specified by a geotechnical engineer.

The next step in sizing the building footings is to determine the materials that will be used. In terms of materials, the IRC provides provisions for the use of cast-in-place concrete, precast concrete, grouted masonry, or wood foundations. Should the designer desire to use another type of material, it would need to be designed under the IBC. The examples used in this section will assume that both the footings and foundations consist of cast-in-place concrete. The reader should review Chapter 4 of the IRC for footing and foundation requirements using other materials.

To size the footings, the designer must use Table R403.1(1) of the IRC. Figure 8.13 provides a simple snapshot of this table. In reviewing the figure, one can see that the left-most column

MINIMUM WIDTH AND THICKNESS FOR CONCRETE FOOTINGS FOR LIGHT-FRAME CONSTRUCTION (inches)[a, b]							
SNOW LOAD OR ROOF LIVE LOAD	STORY AND TYPE OF STRUCTURE WITH LIGHT FRAME	LOAD-BEARING VALUE OF SOIL (psf)					
		1500	2000	2500	3000	3500	4000
20 psf	1 story—slab-on-grade	12 × 6	12 × 6	12 × 6	12 × 6	12 × 6	12 × 6
	1 story—with crawl space	12 × 6	12 × 6	12 × 6	12 × 6	12 × 6	12 × 6
	1 story—plus basement	18 × 6	14 × 6	12 × 6	12 × 6	12 × 6	12 × 6
	2 story—slab-on-grade	12 × 6	12 × 6	12 × 6	12 × 6	12 × 6	12 × 6
	2 story—with crawl space	16 × 6	12 × 6	12 × 6	12 × 6	12 × 6	12 × 6
	2 story—plus basement	22 × 6	16 × 6	13 × 6	12 × 6	12 × 6	12 × 6
	3 story—slab-on-grade	14 × 6	12 × 6	12 × 6	12 × 6	12 × 6	12 × 6
	3 story—with crawl space	19 × 6	14 × 6	12 × 6	12 × 6	12 × 6	12 × 6
	3 story—plus basement	25 × 8	19 × 6	15 × 6	13 × 6	12 × 6	12 × 6
30 psf	1 story—slab-on-grade	12 × 6	12 × 6	12 × 6	12 × 6	12 × 6	12 × 6
	1 story—with crawl space	13 × 6	12 × 6	12 × 6	12 × 6	12 × 6	12 × 6
	1 story—plus basement	19 × 6	14 × 6	12 × 6	12 × 6	12 × 6	12 × 6
	2 story—slab-on-grade	12 × 6	12 × 6	12 × 6	12 × 6	12 × 6	12 × 6
	2 story—with crawl space	17 × 6	13 × 6	12 × 6	12 × 6	12 × 6	12 × 6
	2 story—plus basement	23 × 6	17 × 6	14 × 6	12 × 6	12 × 6	12 × 6
	3 story—slab-on-grade	15 × 6	12 × 6	12 × 6	12 × 6	12 × 6	12 × 6
	3 story—with crawl space	20 × 6	15 × 6	12 × 6	12 × 6	12 × 6	12 × 6
	3 story—plus basement	26 × 8	20 × 6	16 × 6	13 × 6	12 × 6	12 × 6

Figure 8.13 Snapshot of Table R403.1(1) of the IRC. ***(Excerpted from the 2018 International Residential Code; copyright 2017. Washington, D.C.: International Code Council.)***

lists either the roof live load or the ground snow load at the site, whichever governs. The second column notes the size and type of the structure. As discussed earlier, the IRC cannot be used for buildings taller than three stories, so this column allows the designer to select either a one-, two-, or three-story structure. The designer must also determine if the structure will consist of a slab-on-grade, crawl space, or basement. The final step to sizing the footings is to determine the allowable soil-bearing pressure.

The highlighted portions of Fig. 8.13 show how the footings were sized for the following example. A developer is moving a two-story residence from one site to another to make room for their proposed development. The residence will be placed on new concrete footings and stem walls, creating a crawl space. The site is governed by a ground snow load of 30 lb/ft^2 and does not have questionable soils. The jurisdiction requires that a maximum 1500 lb/ft^2 bearing pressure be considered unless a geotechnical report is provided. For this scenario, Table R403.1(1) of the IRC would require a minimum footing width of 17 in. and footing thickness of 6 in. as shown in Fig. 8.13.

Section R403.1.4 of the IRC also requires that the footing be placed at a minimum depth of 12 in. below grade. In areas that are susceptible to frost heave, the footings must be placed below the frost line. As an example, the City of Lincoln, Nebraska, specifies a minimum frost depth for footings of 36 in. (City of Lincoln, n.d.). Rather than extending the footings to below the frost level, the IRC allows the following exceptions:

- The construction of shallow frost-protected foundations that is in accordance with Section R403.3 of the IRC or ASCE 32, *Design and Construction of Frost-Protected Shallow Foundations.* To use shallow frost-protected foundations, the structure must be heated and maintained at a minimum of 64°F. The most important element of these types of foundations is that rigid insulation is provided for a specified distance away from the footing and is also provided vertically up the foundation. This insulation must be protected, and clean gravel must be placed below both the horizontal insulation and the footing.
- The footings can be constructed on solid rock. One should take caution with using this exception, as there are areas of the United States where rock is very porous and is susceptible to frost. As an example, areas of Wasatch County, Utah, contain pot rock deposits (calcareous tufa) that are caused by thermal springs. In such areas, the rock must be removed down to frost depth by means of blasting, or frost-protected foundations must be used.
- Freestanding accessory structures, such as detached garages or sheds, can be constructed on foundations that do not extend down to the frost level. This does not apply to light-framed construction that is larger than 600 ft^2 or concrete or masonry construction that is larger than 400 ft^2. Having a relocatable garage or shed brought to a site is quite common, so this exception could be used often for relocated buildings.

Once the footing size and depth have been selected, it is important to determine the requirements for the concrete stem wall. Remember that in the example used previously, the two-story residence will be placed on concrete stem walls. To determine the stem wall requirements, the designer will need to know a bit more information. The new project site is located within a high-seismic region (Seismic Design Category D2) and requires that the footings be placed at a 30-in. frost depth. To create the crawl space the stem walls should extend 2 ft above the finished grade. After subtracting the thickness of the footing, the overall height of the stem wall would be a total of 4 ft.

Because the height will be no more than 4 ft and it will not retain a significant amount of backfill, it can be designed per Section R403.1.3.1 of the IRC rather than following the requirements for basement foundation walls. Section R403.1.3.1 is meant for stem walls in high-seismic regions and requires that one piece of reinforcing steel be placed within the upper 12 in. of the stem wall and then another at the bottom of the footing, directly below the wall. Because the stem wall concrete is not cast at the same time as the footing concrete, steel reinforcement dowels are required to tie the footing to the foundation. These dowels should consist of reinforcing steel that is placed at no more than 48 in. on center and extends a minimum of 14 in. into the stem wall. Figure 8.14 provides a snapshot of Figure R403.1.3 of the IRC showing how the stem wall should be constructed.

Now that the footing and stem wall have been designed, it is important to determine how the relocated building is to be anchored to the foundation. In the example, an existing two-story wood-framed residence is being anchored to a concrete stem wall. The owner should verify that the existing wood sill plate of the building is made of naturally durable lumber or is preservative treated as required by Section R317.1 of the IRC. If an untreated plate exists, the designer will need to either replace the plate or separate it from the foundation by means of an impervious moisture barrier as discussed in this section. This must be determined prior to anchoring the building to the foundation.

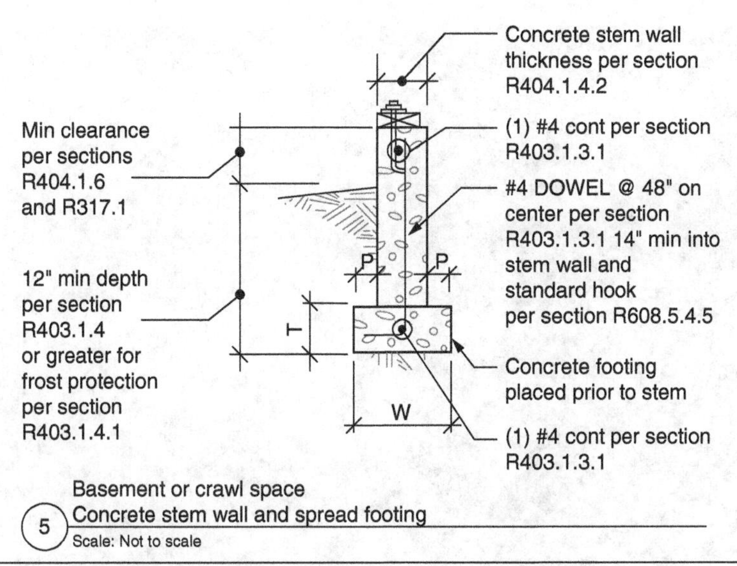

Figure 8.14 Snapshot of Figure R403.1.3 of the IRC. (*Excerpted from the 2018 International Residential Code; copyright 2017. Washington, D.C.: International Code Council.*)

In accordance with Section R403.1.6 of the IRC, a minimum of ½ in. diameter sill anchors should be used having a minimum embedment of 7 in. into the foundation. The bolts should be placed within the middle third of the sill plate, and a minimum of two bolts should be placed in each plate. The anchors should be placed at a spacing of no more than 6 ft but cannot be farther than 12 in. or closer than 4 in. from the ends of the plate. While a standard cut washer and nut can be used in many areas, this example occurs in a high-seismic region, which requires the use of 3-in. by 3-in. by 0.229-in. plate washers.

Another consideration in relation to anchorage is whether mechanical holdowns will be required. The IRC is based on the idea that braced walls will be required. Typically, braced walls are walls with wood sheathing and minimum nailing. In general, they should have minimum widths of 4 ft, although Table R602.10.5 of the IRC may specify lesser or even greater widths in some instances. If the relocated building has several areas with walls that are narrower than 4 ft, the code official may want the design professional to verify that mechanical holdowns are not required. These holdowns are meant to resist any uplift that the building might have at the foundation. Figure 8.15 displays a mechanical holdown at the corner of the building but also shows the sill anchorage that includes the 3-in. by 3-in. plate washers.

In summary, the IRC foundation provisions assist the designer and the code official to determine the requirements for the relocated building's footing, foundation, and anchorage to the foundation. In the example presented, the two-story residence would need to be supported by footings having a width of 17 in., having a thickness of 6 in., and placed 30 in. below grade due to frost. The concrete stem walls will have a minimum thickness of 6 in. and will include a horizontal bar at the top and will have vertical dowels into the footing spaced at 48 in. The building will be anchored to the foundation by means of sill anchors embedded a minimum of 7 in. and spaced no more than 6 ft apart. The designer will also need to determine if mechanical holdowns would be required.

Please note that the discussion above does not cover all the foundation and anchorage provisions of the IRC. Not only is this discussion limited to concrete footings and stem walls but

Figure 8.15 Foundation holdown and sill anchors.

several other topics such as foundation drainage, site drainage, backfill requirements, vapor retarders, foundation vents, and more are not discussed. The reader is referred to Chapter 4 of the IRC should more information be desired.

8.4.2 ***International Building Code.*** When the building being relocated does not fall under the purview of the IRC, the foundation needs to be designed in accordance with the requirements of the IBC. While the IBC does include several prescriptive requirements, it does not provide the specific design for the footings and foundations. Rather, the IBC lays out specific requirements that the structural engineer of record (SER) will need to meet when performing the design of the foundation system and anchorage of the relocated building to the new foundation.

The IRC limits the foundation to shallow systems, while the IBC allows for deep foundations as well. A shallow foundation is defined in Section 202 of the IBC as a continuous wall footing, an isolated spot or mat footing, or a slab-on-grade foundation, while deep foundations are defined simply as a foundation element that does not meet the definition of shallow. Deep foundation systems could be quite common when moving or relocating a building, so the requirements for both will be discussed.

Regardless of the foundation type selected by the SER, they will need to consider several items in their analysis. As was done using the tables in the IRC, the footings will need to be sized based upon the allowable bearing capacity of the soils. In addition, Section 1808.2 of the IBC requires that the design limit differential settlement across the foundation, and Section 1604.8.1 requires that the design also consider sliding and overturning forces at the footing. While the IBC prescribes three different sets of load combinations that can be used when performing structural analysis, Section 1801.2 states that the footing and foundation analysis must consider the allowable stress load combinations outlined in Section 1605.3 of the IBC.

When it comes to the design of foundation walls, they must comply with both Sections 1610 and 1807 of the IBC. These sections require walls that retain soil to be designed for applicable lateral soil pressures. Figure 8.16 outlines some of the loads that a foundation wall might see.

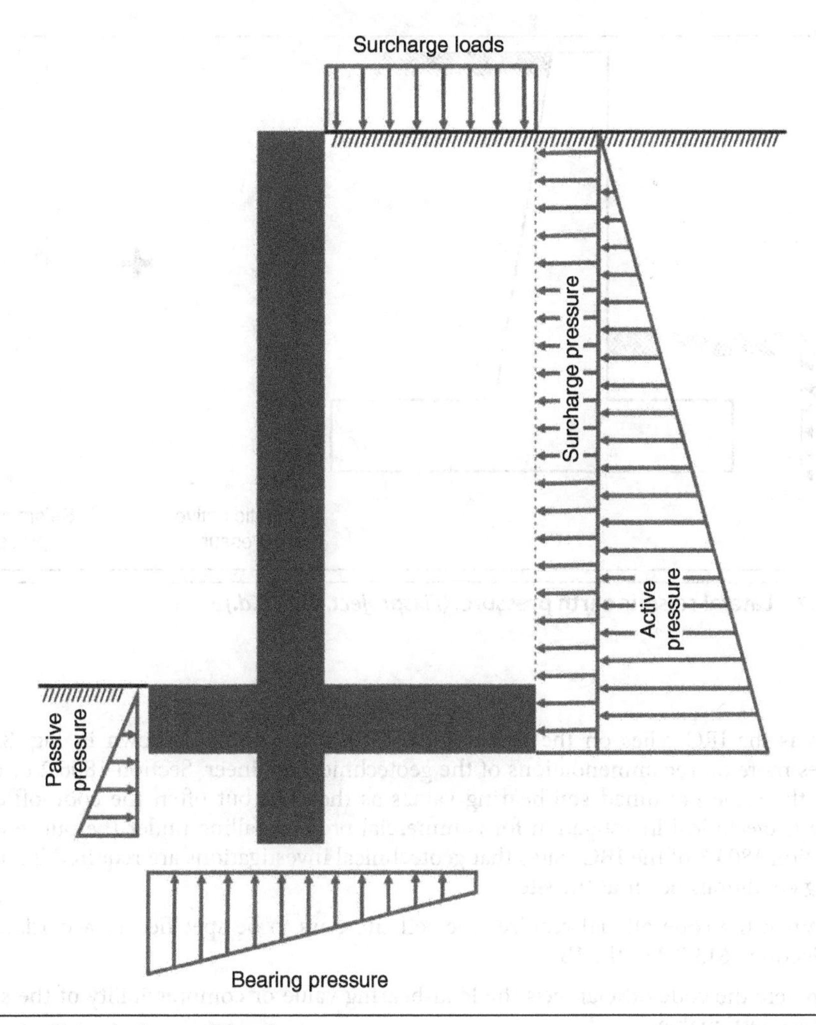

Figure 8.16 Foundation wall loads.

If the foundation wall is restrained at the top by a rigid, or semirigid, floor diaphragm, it can be considered restrained, and the design would not have to consider uplift or sliding loads. Figure 8.16 shows a foundation wall that is not restrained. It also shows a surcharge load at grade near the top of the wall that could be from a parked car, a swimming pool, or other load in that area.

Section 1807 requires that the analysis for unrestrained foundation walls consider a factor of safety of 1.5 against sliding and overturning. Section 1807.2.3 of the IBC allows this safety factor to be reduced to 1.1 when also considering seismic earth pressure. The seismic earth pressure is only required in high-seismic regions where the foundation wall retains more than 6 ft of soil. Section 1803.5.12 of the IBC requires that the geotechnical engineer specify the seismic earth pressure when applicable. There is some debate in the geotechnical community as to how this seismic force is distributed to the foundation wall, but Fig. 8.17 shows the most common method. The reader can note that the seismic pressure is applied as an inverted triangle, while the static soil pressure is applied as an upright triangle.

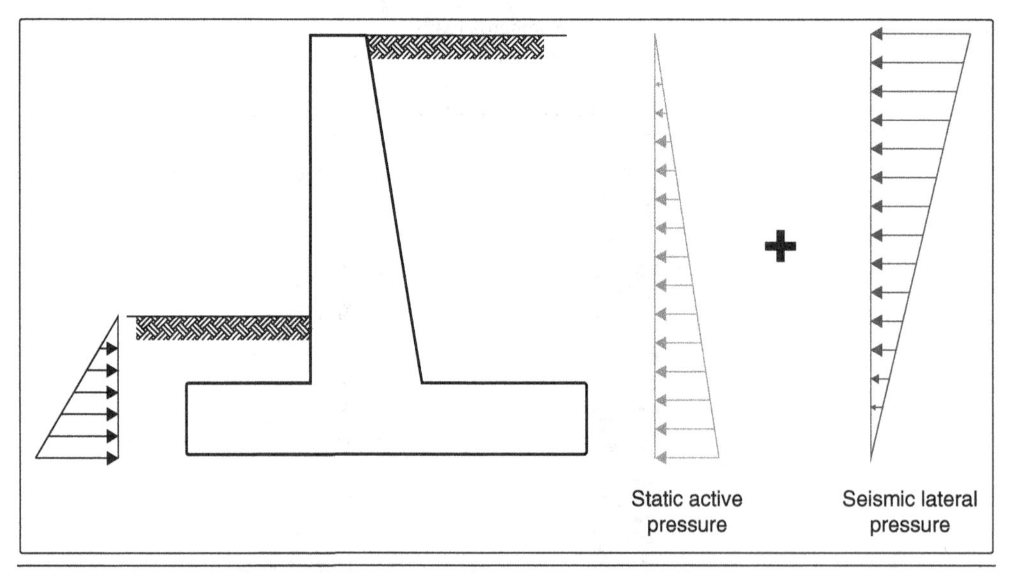

Figure 8.17 Lateral seismic earth pressure. ***(Fixsproject.com, n.d.)***

Whereas the IRC relies on the presumptive load-bearing values shown in Fig. 8.12, the IBC relies more on recommendations of the geotechnical engineer. Section 1806.2 of the IBC includes the same presumed soil-bearing values as the IRC, but often the code official will require a geotechnical investigation for commercial projects falling under the purview of the IBC. Section 1803.5 of the IBC states that geotechnical investigations are required if any of the following conditions occur at the site:

- When the code official requires the soil site class to be specified in accordance with Section 1613.2.2 of the IBC.
- Where the code official feels the load-bearing value or compressibility of the site soils are questionable.
- When site soils have the potential to be expansive.
- When it is determined that the groundwater table might be within 5 ft below the lowest floor level.
- Any time deep foundations are used. The geotechnical engineer must provide recommendations as to type, capacities, spacing, installation procedures, and load testing requirements.
- Where the site has variations in the structure of rock. The most common example would be on a slope where a portion of the building would be founded on rock with the remainder being founded on fill soils.
- Any time that construction will require excavations near existing foundations that could remove the supporting soil. A common rule of thumb is drawing a 45° angle from the edge of the footing down. If any of the soil below the angle is to be removed, the footing may be undermined.

- If the design calls for the footings to be supported on compacted structural fill of 12 in. The geotechnical report must state the type of fill soils to be used, the maximum lift thickness, the minimum in-place density, and the testing requirements.
- While not common, any time controlled low-strength material (CLSM) will be used to support the foundation. This is often termed "flowable fill" and is more often used as backfill material of underground piping, which would not require a geotechnical report.
- The IBC has the same provisions for footings near ascending or descending slopes as was discussed with the IRC. If the owner chooses to have a geotechnical engineer specify alternate setbacks and clearances, they would need to provide a geotechnical report supporting the recommendation to the code official.
- In moderate-seismic regions (i.e., Seismic Design Category C), the geotechnical investigation must address slope instability, liquefaction, total and differential settlement, and the potential for surface displacement due to faulting. Liquefaction is a phenomenon that typically occurs in layers of loose sands that are below the groundwater table. Seismic motions cause these soils to liquefy, which in turn can cause excessive settlements, flotation of buried structures, or even the site soils to spread across the surface like a wave (i.e., lateral spread).
- In high-seismic regions (i.e., Seismic Design Category D or above), there are two main items that need to be addressed in the geotechnical report. The first is listing the seismic lateral earth pressure that should be considered in the design of foundation walls as discussed above. The second is to assess the potential for liquefaction at the site, to list the consequences of the liquefaction-induced hazards, and to list how those consequences are to be mitigated.

For most commercial projects one of the above-listed items will exist at the site, therefore requiring a geotechnical report. Section 1803.6 lists what items must be included in the report. It is quite common for geotechnical reports to only provide a portion of what is listed in Section 1803.6, so it behooves both the design professional and code official to ensure that all required items are provided so the foundation can be designed correctly.

In addition to often requiring a geotechnical investigation, footings and foundations using the IBC often require third-party special inspections to be provided. These inspections are performed by an individual certified to review the specific material requirements and often require the special inspector to be at the site for longer periods of time than the code official. Concrete special inspections often require testing as well to ensure that the concrete meets the design compressive strengths, has adequate air content, and does not exceed the workability limitations of the code.

Section 1705.3 of the IBC includes several special inspection exemptions for continuous and isolated footings, but most concrete foundation walls still require special inspections, as do deep foundation elements. As such, the construction documents for relocated or moved buildings will likely require a Statement of Special Inspections, in accordance with Section 1704.3 of the IBC, to list the special inspections that will be required for the new foundation elements.

The IBC has several other foundation provisions that go above and beyond what is specified in the IRC. This is likely because the IRC mainly deals with simpler residential buildings of light-frame construction, whereas the IBC must be able to address as many conditions as possible. The following two examples will highlight a couple of the additional requirements that are included in the IBC:

- *Footing Seismic Ties:* When the building is being relocated to a high-seismic region and soft or liquefiable soils exist, all footings must be interconnected as shown in Fig 8.18. Section 1809.13 specifically requires this when the project is located within Seismic

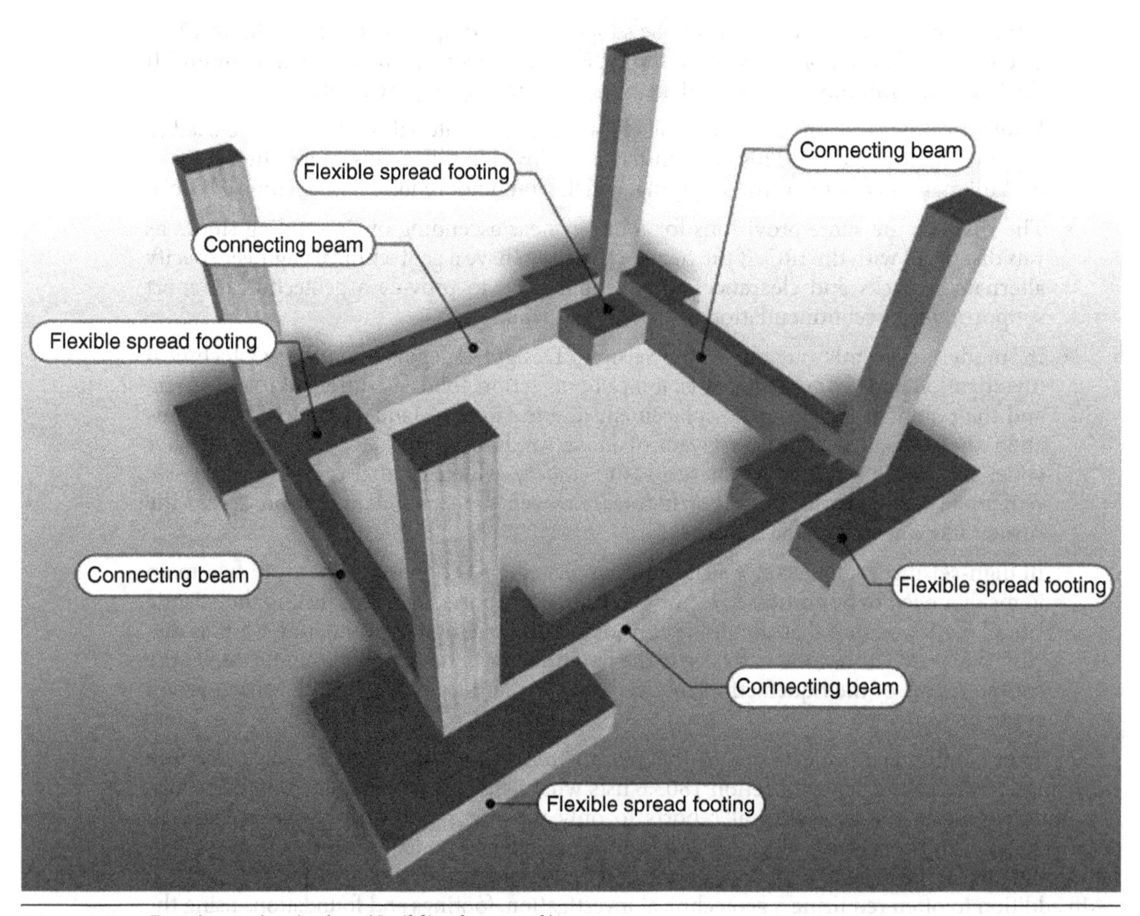

Figure 8.18 Footing seismic ties. (*Buildinghow, n.d.*)

Design Category D or above and the site soils are classified as E or F. Section 20.3.2 of ASCE 7 (ASCE, 2016) defines Site Class E soils as soft clays that are at least 10 ft thick. Section 20.3.1 of ASCE 7 defines Site Class F as soils that are vulnerable to failure and sites with liquefiable soils and peats and organic soils as examples. By interconnecting the footings, the building is better tied together and able to withstand greater amounts of differential settlements that could occur in the poor soils during a seismic event. Similar seismic ties are often required for deep foundation elements as outlined in Section 1810.3.13 of the IBC. If a moved building will require isolated footings and will be located in a high-seismic region with poor soils, it could require the individual footings to be interconnected as outlined in Section 1809.13 of the IBC.

- *Grade Beams:* As the name would suggest, a grade beam is simply a concrete beam that rests on the soil, or grade. While it is surrounded by soil, it is required to span a greater distance than a standard continuous footing and often is required to take axial compression or tension loads. Figure 8.18 shows grade beams being used to provide the footing seismic ties discussed previously. Section 1810.3.12 of the IBC notes that grade beams in high-seismic regions are to be designed in accordance with Chapter 14 of ACI 318-14.

This chapter requires the grade beams to have both top and bottom continuous reinforcement and enclosed ties at minimum spacing throughout the length of the beam (ACI, 2014). It is possible that a relocated building could be placed on continuous footings that are designed and detailed as grade beams.

While the focus thus far has been on the foundation requirements, it is important to remember that Section 1402.2.1 of the IEBC requires that the moved building be properly anchored to the new foundation. Though the IRC specified exactly how the building was to be anchored, the IBC requires that the SER perform an analysis and then provide details on the construction documents. Section 1604.8 of the IBC requires that the "required anchorage load" be calculated using ASCE 7-16, *Minimum Design Loads and Associated Criteria for Buildings and Other Structures,* as published by the American Society of Civil Engineers (ASCE). As most anchorages occur in concrete foundations, Section 1901.3 of the IBC requires that the "allowable anchorage load" be determined using ACI 318, *Building Code Requirements for Structural Concrete,* as developed by the American Concrete Institute (ACI). The analysis should show that the "allowable" load exceeds the "required" load determined in the calculations.

In summary, when designing a new foundation system for a moved building in accordance with the IBC, the SER will need to perform a detailed analysis considering allowable bearing pressure, lateral soil pressures, differential settlements, overturning loads, and sliding loads. The analysis for anchoring the building to the foundation should be performed in accordance with Chapter 17 of ACI 318-14. Before a building permit can be issued for the work, the code official will likely require structural calculations, a geotechnical investigation, and a Statement of Special Inspections in addition to the construction documents.

Please note that this discussion of the foundation and anchorage design using the IBC is general in nature. Other considerations will also need to be made above and beyond what has been discussed here, including excavation and grading requirements, damp proofing and waterproofing, concrete cover, and much more. Please refer to Chapters 16, 18, and 19 of the IBC for more information in relation to foundation design.

8.5 *Wind, Seismic, Snow, and Flood Loads*

Relocated or moved buildings are not required to meet all the structural design requirements prescribed by the IBC or IRC, but they are required to meet certain wind, seismic, snow, and flood load requirements. The following subsections outline the IEBC minimum requirements in relation to each of these structural loads.

8.5.1 Wind loads. Section 1402.3 of the IEBC requires that relocated or moved buildings meet the wind provisions specified in either the IBC or IRC. In both the IBC and IRC, the first step to checking compliance with the wind provisions is to determine what the design wind speed and wind exposure category at the new site should be.

The wind speed is determined by referring to maps of the United States, which are provided in both the IBC and IRC. Significant changes have occurred in these maps from the 2015 version of the code to the current version. For comparison purposes, Fig. 8.19 displays both the 2015 and 2018 versions of Figure 1609.3(1) in the IBC. This figure is for Risk Category II structures only and mirrors the map provided in the IRC.

When comparing the two versions of the map in Fig. 8.19, one can see that wind speeds have generally decreased across the United States. The exception to this would be along the coastline between Texas and North Carolina. The Western states will see the most significant reduction, on the order of 15 percent. The changes that have been made to the wind speed maps are based upon a study of wind data at over 1000 recording stations across the country (Scott, 2018.)

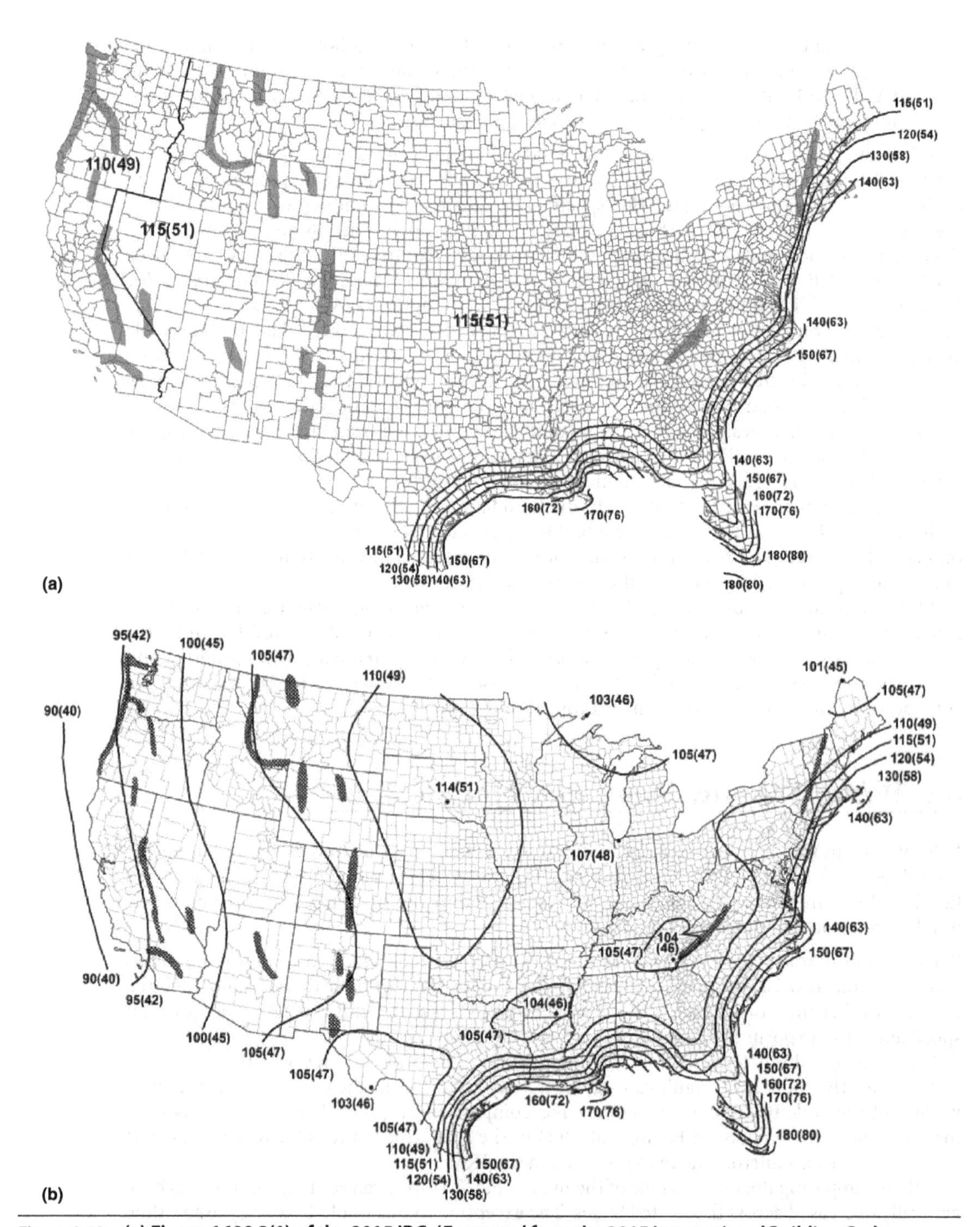

Figure 8.19 (*a*) Figure 1609.3(1) of the 2015 IBC. (*Excerpted from the 2015 International Building Code; copyright 2014. Washington, D.C.: International Code Council.*). (*b*) Figure 1609.3(1) of the 2018 IBC. (*Excerpted from the 2018 International Building Code; copyright 2017. Washington, D.C.: International Code Council.*)

After determining the design wind speed, the designer must select the appropriate wind exposure category for the new site. The wind exposure category is based upon the relocated building's surroundings that might slow down the wind. This includes items such as other buildings, trees, or hills. The code provides extra emphasis on obstructions that are at least 30 ft in height, or roughly the size of an average home.

Figure 8.20 provides an example of the three wind exposure categories defined in Section 1609.4.3 of the IBC (see also R301.2.1.4 of the IRC). Locations meeting the requirements

Wind Exposure Category B
Suburban neighborhood in Bloomington, Illinois
Wind Exposure Category C
Are in Nashville, Tennessee, that is generally open terrain with scattered trees
Wind Exposure Category D
Homes along coast of Lake Superior near Silver Bay, Minnesota

Figure 8.20 Wind exposure categories.

of Wind Exposure Category B have closely spaced obstructions having a height of at least 30 ft. To be classified as Category B these obstructions should extend around the relocated building for a distance of 1500 ft in all directions. Wind Exposure Category C is selected when the site has some obstructions, but they are not close enough together or they do not extend for 1500 ft in all directions. Wind Exposure Category D includes those areas where the surface around the relocated building is flat or unobstructed for a distance of at least 5000 ft. This could occur next to water surfaces, as shown in Fig. 8.20, or mud flats, salt flats, or large bodies of ice.

Most jurisdictions only have areas that could be classified as either Category B or C. The wind pressures that are calculated using Category C are significantly greater than those determined using Category B. As such, the code official should verify that the designer has considered an appropriate wind exposure category. Tools such as Google Earth are extremely helpful when making this determination.

Whether following the path of the IBC or the IRC, the first two steps are to determine the wind speed and wind exposure. After those have been determined, the two code paths separate. To check compliance with the IRC, the designer should first determine whether the components of the relocated building have been attached in compliance with Section R602.3 of the IRC. While Table R602.3(1) details the requirements for 39 separate structural member connections, the Federal Emergency Management Agency (FEMA) emphasizes that homes subject to high winds should focus on load path connections and the attachment of roof and wall coverings (FEMA P-804, 2010).

In relation to the load path, it is important the owner or evaluator of the existing building ensure that loads can get to the foundation. Whether those loads originate at the roof or at the floor level, they must be able to find their way to the foundations. While it is important that structural members are appropriately sized in relation to the load path, it is more important that appropriate connections are provided to transfer the loads from one structural member into the next. Figure 8.21 highlights three key connections in terms of the wind load path.

The first is the roof-to-wall connection. This is key to ensure that the trusses or roof framing is attached sufficiently to resist uplift forces at the eave but also to transfer gravity forces into the wall framing and lateral loads into the wall sheathing. The second connection occurs between the walls and the elevated floors. Adequate connections need to be provided to again ensure that gravity and lateral loads can be transferred, and often metal straps are required to transfer uplift forces from one level to the next. The last connection is attaching the wall framing to the foundation. This requires that appropriate sill anchors and spacings are provided and that mechanical holdowns are placed to resist uplift forces as needed.

When evaluating the adequacy of the wind load path in a relocated building, the evaluator must be able to answer the following questions:

- Are roof-to-wall connections able to resist wind uplift and shear loads?
- For multistory homes, are floor-to-wall connections able to resist wind cumulative wind uplift as well as wind shear loads?
- Are wall-to-foundation connections capable of resisting cumulative wind uplift and wind shear loads?
- Is a complete load path provided at attached structures (e.g., carports, covered patios, decks, etc.)?

Besides the load path connections, FEMA P-804 also highlights the need for proper attachment of roof and wall coverings. In relation to roof coverings, it is important to ensure that the roof sheathing is attached adequately to the roof framing members, and then that the roof covering itself (e.g., asphalt shingles, clay or concrete tiles, etc.) is appropriate.

In relation to the roof sheathing, Table R602.3(1) of the IRC requires that it be attached to roof framing members using a minimum of 6d common nails spaced at 6 in. along the edges of the sheathing and at 12 in. on center at intermediate framing members. FEMA recommends

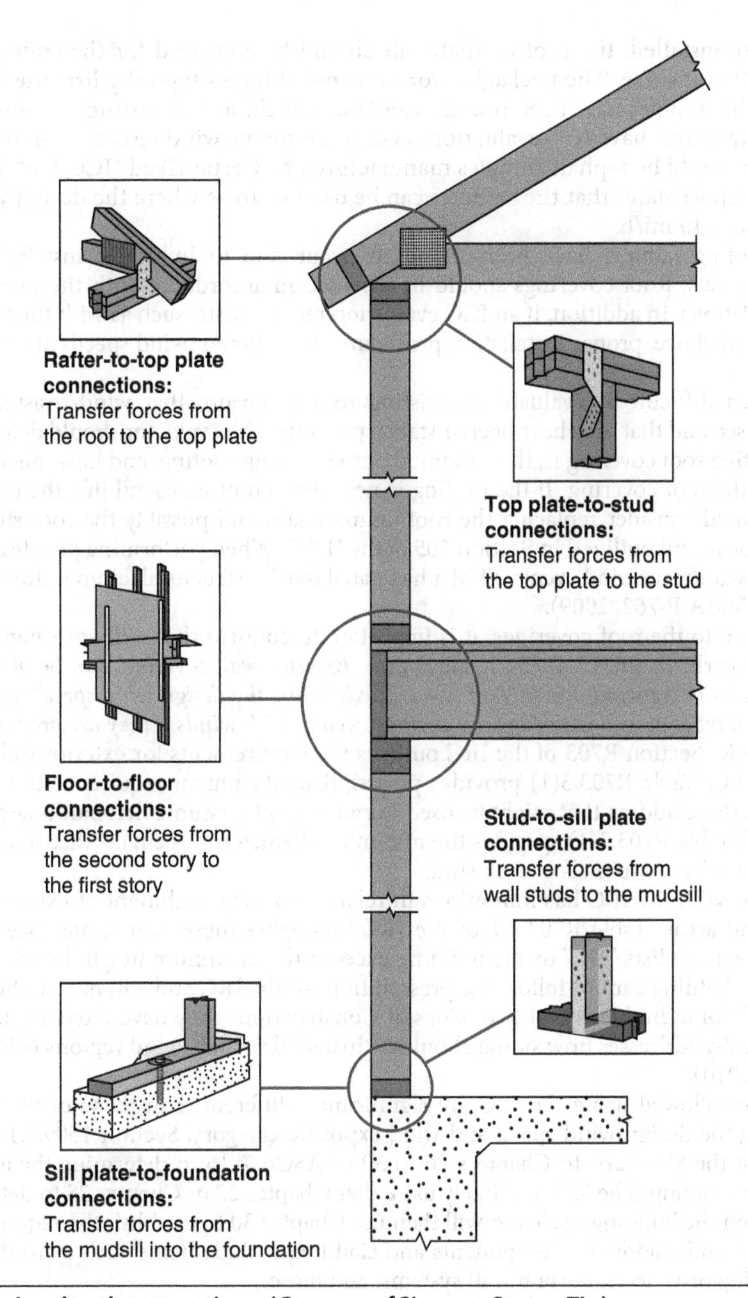
Rafter-to-top plate connections:
Transfer forces from the roof to the top plate
Top plate-to-stud connections:
Transfer forces from the top plate to the stud
Floor-to-floor connections:
Transfer forces from the second story to the first story
Stud-to-sill plate connections:
Transfer forces from wall studs to the mudsill
Sill plate-to-foundation connections:
Transfer forces from the mudsill into the foundation

Figure 8.21 Load path connections. (*Courtesy of Simpson Strong-Tie.*)

that the nailing at intermediate framing members also be at 6 in. on center where the wind speed is 120 mi/h or less and recommends nail spacings of 4 in. on center where the wind speeds are greater (FEMA P-804, 2010).

With respect to roof coverings, the evaluator should consider if the appropriate products have been used and if they have been installed correctly. To ensure that appropriate products are installed, the roofing materials should be approved for the code-prescribed wind speeds at the site. The packaging for new roof shingles typically lists the wind resistance that the shingles have been tested to and likely include a UL listing. In addition, most large manufacturers have ICC evaluation reports noting the wind resistance of the shingles. An example would be asphalt shingles manufactured by CertainTeed (ICC ESR-1389). This evaluation report states that the shingles can be used in areas where the design wind speed is as much as 140 mi/h.

Most roofing failures have been found to occur due to improper installation of the roofing materials. Roof coverings should be installed in accordance with the manufacturer's recommendations. In addition, if an ICC evaluation report exists, such as with the CertainTeed product noted above, proper installation procedures for different wind speeds are listed within the report.

It is often difficult to evaluate an existing roof to ensure that wind-resistant shingles have been used and that they have been installed properly. The evaluator should determine how old the existing roof covering is, the condition of the existing roofing, and how much useful life remains in the roof covering. If the roofing is near the end of its useful life, the owner of the building should consider replacing the roofing materials, and possibly the roof sheathing, as discussed above and outlined in Section 705 of the IEBC. When performing postdisaster evaluations after significant wind events, FEMA has stated that "...structural failures often begin with the roof" (FEMA P-762, 2009).

In addition to the roof coverings, it is important to confirm that wall coverings have been attached properly. In FEMA P-804 it states that *"Exterior wall coverings can be blown off of a building, even during wind events with wind speeds below the design wind speed."* It goes on to state that *"All types of wall coverings can perform well in high winds if they are properly installed for high winds."* Section R703 of the IRC outlines the requirements for exterior wall coverings under the IRC. Table R703.3(1) provides prescriptive attachment requirements for different types of exterior cladding that might be used. As an example, assume that hardboard lap siding will be used. Table R703.3(1) specifies the minimum diameter of the nails that should be used and requires at least two nails at each stud.

With that said, the IRC has limitations in relation to the attachment of exterior cladding in high-wind areas. Table R703.3.1 of the IRC highlights these limitations (see Fig. 8.22). Any time the table lists "DR" or the building exceeds the maximum height listed, the attachment of the cladding cannot follow the prescriptions of the IRC and will need to be designed. In FEMA's "Home Builder's Guide to Coastal Construction" they have a technical fact sheet that specifically addresses how siding should be installed in high-wind regions (FEMA P-499, December 2010).

The steps followed using the IBC are significantly different than those of the IRC. After determining the design wind speed and wind exposure category, Section 1609.1.1 of the IBC requires that the SER turn to Chapters 26 to 30 of ASCE 7-16 to determine the actual wind loads on the building. The SER will likely use either Chapter 27 or Chapter 28 to determine the wind loads on the building itself and will then use Chapter 30 to establish the requirements for "components and cladding." Components and cladding include items such as roof coverings, exterior siding or veneers, curtain wall systems, and more.

After determining the wind loads, the SER should evaluate the existing building and determine whether it conforms or if there are items that will need corrections. The same concepts discussed under the IRC also apply to projects using the IBC. Is a complete wind load path provided addressing both uplift and shear loads? Are roof and wall coverings appropriate for the calculated wind loads, and have they been installed correctly? Are projections (e.g., canopies, balconies, etc.) appropriately attached?

SIDING MINIMUM ATTACHMENT AND MINIMUM THICKNESS									
				TYPE OF SUPPORTS FOR THE SIDING MATERIAL AND FASTENERS					
SIDING MATERIAL		NOMINAL THICKNESS (inches)	JOINT TREATMENT	Wood or wood structural panel sheathing into stud	Fiberboard sheathing into stud	Gypsum sheathing into stud	Foam plastic sheathing into stud[l]	Direct to studs	Number or spacing of fasteners
Anchored veneer: brick, concrete, masonry or stone (see Section R703.8)		2	Section R703.8	Section R703.8					
Adhered veneer: concrete, stone or masonry (see Section R703.12)		—	Section R703.12	Section R703.12					
Fiber cement siding	Panel siding (see Section R703.10.1)	5/16	Section R703.10.1	6d common (2" × 0.113")	6d common (2" × 0.113")	6d common (2" × 0.113")	6d common (2" × 0.113")	4d common (1½ " × 0.099")	6" panel edges 12" inter. sup.
	Lap siding (see Section R703.10.2)	5/16	Section R703.10.2	6d common (2" × 0.113")	6d common (2" × 0.113")	6d common (2" × 0.113")	6d common (2" × 0.113")	6d common (2" × 0.113") or 11 gage roofing nail	Note f
Hardboard panel siding (see Section R703.5)		7/16	—	0.120" nail (shank) with 0.225" head	0.120" nail (shank) with 0.225" head	0.120" nail (shank) with 0.225" head	0.120" nail (shank) with 0.225" head	0.120" nail (shank) with 0.225" head	6" panel edges 12" inter. sup.[d]
Hardboard lap siding[j] (see Section R703.5)		7/16	Note e	0.099" nail (shank) with 0.240" head	0.099" nail (shank) with 0.240" head	0.099" nail (shank) with 0.240" head	0.099" nail (shank) with 0.240" head	0.099" nail (shank) with 0.240" head	Same as stud spacing 2 per bearing
Horizontal aluminum[a]	Without insulation	0.019[b]	Lap	Siding nail 1½" × 0.120"	Siding nail 2" × 0.120"	Siding nail 2" × 0.120"	Siding nail[h] 1½" × 0.120"	Not allowed	Same as stud spacing
		0.024	Lap	Siding nail 1½" × 0.120"	Siding nail 2" × 0.120"	Siding nail 2" × 0.120"	Siding nail[h] 1½" × 0.120"	Not allowed	
	With insulation	0.019	Lap	Siding nail 1½ 0.120"	Siding nail 1½" × 0.120"	Siding nail 1½" × 0.120"	Siding nail[h] 1½" × 0.120"	Siding nail 1½" × 0.120"	
Insulated vinyl siding[j]		0.035 (vinyl siding layer only)	Lap	0.120 nail (shank) with a 0.313 head or 16-gage crown[h, i]	0.120 nail (shank) with a 0.313 head or 16-gage crown[h]	0.120 nail (shank) with a 0.313 head or 16-gage crown[h]	0.120 nail (shank) with a 0.313 head Section R703.11.2	Not allowed	16 inches on center or specified by manufacturer instructions, test report or other sections of this code
Particleboard panels		3/8	—	6d box nail (2" × 0.099")	6d box nail (2" × 0.099")	6d box nail (2" × 0.099")	6d box nail (2" × 0.099")	Not allowed	6" panel edges 12" inter. sup.
		1/2	—	6d box nail (2" × 0.099")	6d box nail (2" × 0.099")	6d box nail (2" × 0.099")	6d box nail (2" × 0.099")	6d box nail (2" × 0.099")	
		5/8	—	6d box nail (2" × 0.099")	8d box nail (2½" × 0.1 23")	8d box nail (2½" × 0.113")	6d box nail (2" × 0.099")	6d box nail (2" × 0.099")	
Polypropylene siding[k]		Not applicable	Lap	Section 703.14.1	Section 703.14.1	Section 703.14.1	Section 703.14.1	Not allowed	As specified by the manufacturer instructions, test report or other sections of this code

(continued)

Figure 8.22 Table R703.3.1 of the IRC. ***(Excerpted from the 2018 International Residential Code; copyright 2017. Washington, D.C.: International Code Council.)***

SIDING MINIMUM ATTACHMENT AND MINIMUM THICKNESS									
				TYPE OF SUPPORTS FOR THE SIDING MATERIAL AND FASTENERS					
SIDING MATERIAL		NOMINAL THICKNESS (inches)	JOINT TREATMENT	Wood or wood structural panel sheathing into stud	Fiberboard sheathing into stud	Gypsum sheathing into stud	Foam plastic sheathing into stud[l]	Direct to studs	Number or spacing of fasteners
Steel[c]		29 ga.	Lap	Siding nail (1¾ × 0.113") Staple–1¾"	Siding nail (2¾ × 0.113") Staple–2½"	Siding nail (2½ × 0.113") Staple–2¼"	Siding nail (1¾ × 0.113") Staple–1¾"	Not allowed	Same as stud spacing
Vinyl siding (see Section R703.11)		0.035	Lap	0.120" nail (shank) with a 0.313" head or 16-gage staple with ⅜ - to ½-inch crown[h, i]	0.120" nail (shank) with a 0.313" head or 16-gage staple with ⅜ - to ½ -inch crown[h]	0.120" nail (shank) with a 0.313" head or 16- gage staple with ⅜ - to ½ -inch crown[h]	0.120" nail (shank) with a 0.313 head Section R703.11.2	Not allowed	16 inches on center or as specifiedby the manufacturer instructions ortest report
Wood siding (see Section R703.5)	Wood rustic, drop	3/8 min.	Lap	6d box or siding nail (2" × 0.099")	6d box or siding nail (2" × 0.099")	6d box or siding nail (2" × 0.099")	6d box or siding nail (2" × 0.099")	8d box or siding nail (2½" × 0.113") Staple–2"	Face nailing up to 6" widths, 1 nail per bearing; 8" widths and over, 2 nailsper bearing
	Shiplap	19/32 average	Lap						
	Bevel	7/16							
	Butt tip	3/16	Lap						
Wood structural panel ANSI/APA PRP-210 siding (exterior grade) (see Section R703.5)		3/8–1/2	Note e	2" × 0.099" siding nail	2½" × 0.113" siding nail	2½" × 0.113" siding nail	2½" × 0.113" siding nail	2" × 0.099" siding nail	6" panel edges 12" inter. sup.
Wood structural panellap siding (see Section R703.5)		3/8–1/2	Note e Note g	2" × 0.099" siding nail	2½" × 0.113" siding nail	2½" × 0.113" siding nail	2½" × 0.113" siding nail	2" × 0.099" siding nail	8" along bottom edge

For SI: 1 inch = 25.4 mm.

a. Aluminum nails shall be used to attach aluminum siding.

b. Aluminum (0.019 inch) shall be unbacked only where the maximum panel width is 10 inches and the maximum flat area is 8 inches. The tolerance for aluminum siding shall be +0.002 inch of the nominal dimension.

c. Shall be of approved type.

d. Where used to resist shear forces, the spacing must be 4 inches at panel edges and 8 inches on interior supports.

e. Vertical end joints shall occur at studs and shall be covered with a joint cover or shall be caulked.

f. Face nailing: one 6d common nail through the overlapping planks at each stud. Concealed nailing: one 11-gage 1½-inch-long galv. roofing nail through the top edge of each plank at each stud in accordance with the manufacturer's installation instructions.

g. Vertical joints, if staggered, shall be permitted to be away from studs if applied over wood structural panel sheathing.

h. Minimum fastener length must be sufficient to penetrate sheathing other nailable substrate and framing a total of a minimum of 1½ inches or in accordance with the manufacturer's installation instructions.

i. Where specified by the manufacturer's instructions and supported by a test report, fasteners are permitted to penetrate into or fully through nailable sheathing or other nailable substrate of minimum thickness specified by the instructions or test report, without penetrating into framing.

j. Insulated vinyl siding shall comply with ASTM D7793.

k. Polypropylene siding shall comply with ASTM 7254.

l. Cladding attachment over foam sheathing shall comply with the additional requirements and limitations of Sections R703.15, R703.16 and R703.17.

Figure 8.22 ***(Continued)***

In summary, to show compliance with the wind provisions of the IRC or IBC, the designer must determine the design wind speed, select the appropriate wind exposure category, and then determine if appropriate members and connections have been provided to resist the assessed wind pressures. To show compliance, a report should be provided to the code official noting how the existing building conforms to the current code or note what items need to be brought up to current standards. Should deficiencies exist, construction documents should be provided clearly noting how substandard items will be corrected.

Section 1402.3 of the IEBC does allow for two exceptions that would not require the moved building to meet the wind requirements of the current code. The first of these exceptions is specific to detached one- and two-family dwellings as well as Group U occupancies. This exception does not require a check for wind if the design wind loads at the new site are either equal to or less than the wind loads at the previous site. Many owners, or even code officials, might think this simply means that the wind speed at the new site simply needs to be greater than or equal to the wind speed at the previous site. This is not the case, as both the wind speed and wind exposure category must be considered. Table 8.1 shows when the wind loads are likely equal to or less than at the previous site and when a wind analysis to ensure compliance would be needed.

Table 8.1 Wind Speed and Wind Exposure Category Comparison

	Same Wind Exposure Category	Reduced Exposure Category	Greater Exposure Category
Same wind speed	Okay	Okay	Wind check
Reduced wind speed	Okay	Okay	Wind check
Greater wind speed	Wind check	Wind check	Wind check

The second exception provided in Section 1402.3 is in relation to the *10 percent rule* discussed previously in this book. It states that if the stress in the structural elements has not increased by more than 10 percent as a result of the relocation, the wind check is not required.

8.5.2 Seismic loads. Section 1402.4 of the IEBC requires that the moved building meet the seismic design requirements of the IBC for the new location. There are two exceptions to this rule as highlighted below:

1. If the building will be located within a Seismic Design Category A or B, or if it is a detached one- or two-family building that will be located in Seismic Design Category A, B, or C, such an analysis is not required as long as the seismic ground motions are less than they were at the previous site.
2. The *10 percent lateral rule* still applies. So long as the lateral forces are not increased by more than 10 percent from the previous site, an analysis showing compliance with the IBC will not be required.

8.5.3 Snow loads. When a building is moved or relocated to a site that has higher snow loads, Section 1402.5 of the IEBC requires that it comply with the requirements of either the IBC or IRC. If it is relocated to a site with equal or lesser snow loads, this analysis would not be required. This section also allows the 5 percent gravity rule as an exception.

8.5.4 Flood loads. Section 1402.6 of the IEBC simply requires that buildings that are moved or relocated to a flood hazard area meet the flood-resistant design provisions of Section 1612 in the IBC or Section R322 of the IRC. There are no exceptions to compliance with this requirement. Moving an existing building into a known floodplain goes against the better judgment of most people. In FEMA's "Homeowner's Guide to Retrofitting" they highlight six ways to protect one's home from flooding (FEMA P-312, 2014). Chapter 6 of this guide focuses on relocating the existing building to be out of the flood hazard area, rather than into it.

As an example of what is required by this section, assume that a mobile home will be moved to the City of Albany, Georgia, and will be located within a 100-year floodplain. The City requires that all manufactured homes be placed on permanent foundations that are at least 1 ft above the base flood elevation (BFE). The foundation must be able to resist flotation, collapse, and lateral movement when considering the flood loads. The City also requires that utilities, such as propane tanks, be anchored to elevated platforms that are above the BFE (City of Albany, n.d.). Figure 8.23, developed by the Georgia Department of Natural Resources, portrays the requirements for mobile homes in the City of Albany.

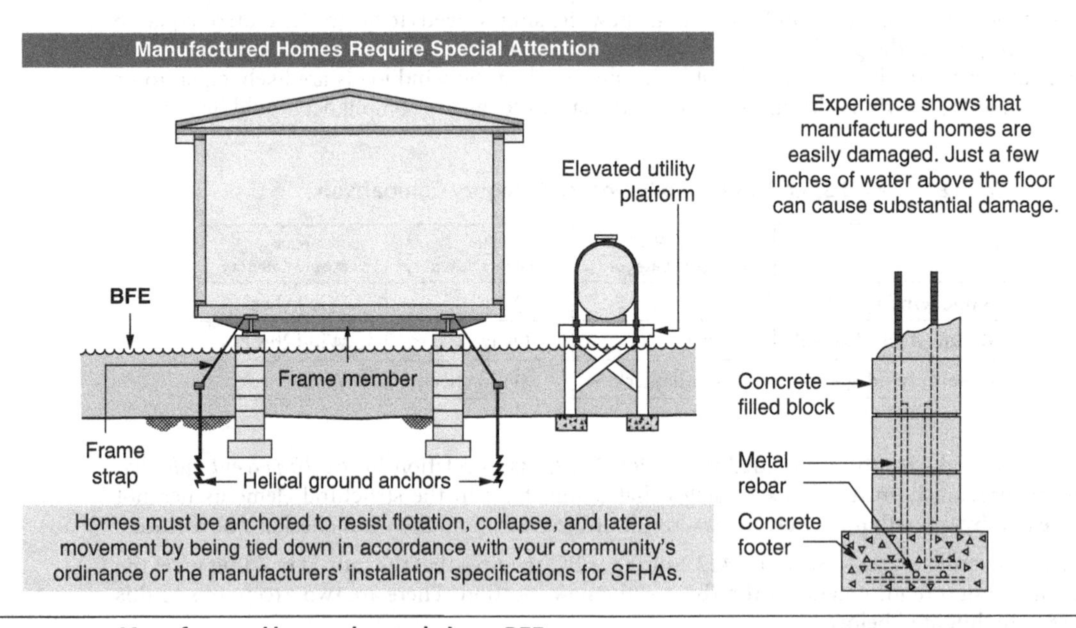

Figure 8.23 Manufactured home elevated above BFE.

8.6 *Inspections and Repairs*

Section 1402.7 of the IEBC allows the code official to require an inspection of the building after it has been moved or relocated. This inspection can either be performed by the code official themselves or they can require "approved" professionals to perform the inspection. Most jurisdictions would likely require that this inspection be performed by a third party rather than taking that responsibility on themselves.

The purpose of this inspection is to examine the structural parts of the building and to ensure that the moving of the structure did not cause any of the components or their connections to sustain any structural damage. If performed by a third party, a report should be presented to the code official noting any items that require correction. The code official would then likely require that appropriate repairs are made before the relocation work is completed and a certificate of occupancy is issued.

8.7 *Conclusion*

In summary, Chapter 14 of the IEBC requires that the moved or relocated building comply with the following items:

- Conforms to the 2018 IFC and IPMC.
- Repairs must be performed in accordance with Chapter 4 of the IEBC.
- Alterations or additions should be performed in accordance with either the prescriptive, work area, or performance compliance method.
- New construction complies with the IBC and IRC.
- It is placed on the lot to comply with the provisions previously discussed in Sec. 8.3.
- Foundations are to be designed in accordance with the IBC and IRC.
- The building must comply with the IBC or IRC wind, seismic, snow, and hazard loads unless one of the exemptions has been met.
- An inspection must be performed after the building has been moved to identify any structural damage that needs to be repaired due to the moving operations.

While the above-noted items are what is required by the IEBC specifically, local jurisdictions may have additional requirements. The following is a listing of some items that might be required as part of moving or relocating a structure, which are not specifically identified in Chapter 14 of the IEBC:

- *Pre-Relocation Inspection:* Prior to moving the building, FEMA recommends that an evaluation be performed to check the age and condition of the home, its overall structural integrity, any weaknesses in the structure or envelope, and whether wind or seismic retrofits would likely be needed (FEMA P-804, December 2010). While this is a good idea to do, some jurisdictions may desire to inspect the structure as well before it is moved. Spokane County, Washington (2011), performs a "pre-relocation" before they issue a permit to verify the condition of the structure and advise the permit applicant as necessary prior to any moving operations.
- *Compatibility Study:* When a building is being moved into the City of Los Angeles, California, a compatibility study is required to be performed (City of Los Angeles, 2014). As part of the study, a city inspector visits the proposed site to ascertain whether the proposed building is compatible with the surrounding neighborhood.
- *Determination of Habitability:* Rather than demolish a structure that is considered habitable, the county of Santa Cruz, California, requires a determination of habitability before a demolition permit is granted (County of Santa Clara, 1983). This determination must be made by the code official within 10 days of a demolition permit application. If a structure is deemed habitable, a public notice must be made stating that the building is available for sale or removal.
- *Restoring Site:* If a building is to be moved from one site to another, most jurisdictions require that the old site be restored to an acceptable level. Some of the items that may be required include the removal of old foundations and pavement, backfilling basements, removing abandoned utilities, site grading, and landscaping (FEMA P-312, 2014).

CHAPTER

9

CONSTRUCTION SAFEGUARDS

9.1 *General*

Chapter 15 of the *International Existing Building Code*® (IEBC®) addresses precautions that should be followed when performing construction on existing buildings. These precautions are needed to ensure the safety of the occupants within the building as well as pedestrians who may be passing near the jobsite or to protect adjacent properties that could be adversely affected by the work being performed.

Section 1501 of the IEBC addresses general precautions that should be made. The following subsections provide a brief description of what should be provided in relation to these subsections.

9.1.1 Fire safety. Fire safety during construction must comply with the requirements of the *International Building Code*® (IBC®) as well as Chapter 33 of the *International Fire Code*® (IFC®). Additional requirements are outlined in Sec. 9.3 of this chapter.

9.1.2 Protection of pedestrians. Pedestrians must be protected during any construction or demolition activities. Such protection could require protected walkways (see Fig. 9.1), directional barricades, construction railings, barriers, signage, and more.

Figure 9.1 Sidewalk protection. (*Systems, n.d.*)

9.1.3 Sanitary facilities. Construction activities must provide sanitary facilities as required by the *International Plumbing Code*® (IPC®).

9.2 *Adjoining Properties*

Section 1502 of the IEBC addresses the need to protect adjoining properties during any construction activities or demolition work. It addresses the need to protect existing structural and fire-resistance-rated members, to control water runoff, and to notify adjoining building owners prior to any excavation work. Chapter 2 of this book notes that several jurisdictions, such as New York City, actually require third-party special inspections of adjoining buildings during the construction activities.

9.3 *Fire and Life Safety Precautions*

Chapter 15 of the IEBC includes several sections that require minimum fire and life safety provisions to be in place during construction. The following is a brief description of several of these items.

9.3.1 Fire extinguishers. Fire extinguishers must be placed at each stairway of all floors that contain combustible materials, in all storage and construction sheds, and in locations where hazardous materials exist within an active construction site. Such extinguishers should be easily located and readily accessed as shown in Fig. 9.2.

Figure 9.2 Construction site fire extinguisher. (*Elcosh, 2014.*)

9.3.2 Means of egress. For existing buildings undergoing construction, the means of egress system must be maintained throughout the construction process unless an "approved" temporary means of egress is provided. In addition, once the construction reaches a height of 40 ft above the lowest level of fire department access, a stairway must be provided. Such stairway must extend to within one floor of the highest point of construction.

9.3.3 Standpipes. Section 905.3.1 of the IBC requires all buildings to have a standpipe system if the highest occupiable floor level is more than 30 ft above the lowest level of fire department access. Section 1506 of the IEBC states that if the IBC requires a building to have a standpipe system, a system must be installed during construction once the overall height of construction exceeds 40 ft above the lowest level of fire department access. The standpipes shall be located adjacent to stairways and must extend to within one floor of the highest point of construction.

9.3.4 Fire sprinklers. Section 1507 of the IEBC requires that fire sprinklers be in place and tested prior to occupying the space. This requirement is only triggered if Section 903 of the IBC would require such sprinkler protection. In addition, if control valves are used to separate spaces, they must be checked at the end of each work period to ensure that fire suppression is active and working.

9.3.5 Fire protection water supply. Prior to any combustible material being brought to the site, an approved water supply must be in place. In order to be approved, the fire protection water supply must typically meet the requirements of Appendix B in the IFC. The required fire flow is dependent upon the type of construction and the size of the building in question.

9.4 *Accessibility*

Section 1508 of the IEBC exempts most elements of a construction project from the accessibility requirements of the code. This includes items such as scaffolding, material hoists, construction trailers, and the like.

CHAPTER

10

IEBC APPENDICES

10.1 *Appendix A: Seismic Retrofit*

Appendix A of the *International Existing Building Code®* (IEBC®) is set aside to provide methods for seismically retrofitting existing buildings. These methods have been around for quite a while in many cases and have a proven track record. As has been discussed previously in this book, the Appendix A seismic retrofit provisions can only be used when the IEBC allows for the lateral evaluation to be performed to the *reduced seismic* level. If *full seismic* is required, the Appendices cannot be used. The purpose of these seismic retrofit procedures is to identify reasonable steps that can be taken to reduce, or eliminate, seismic risk in existing buildings. The following provides a brief description of each of the seismic retrofit options included in Appendix A of the IEBC.

10.1.1 Chapter A1: Unreinforced masonry (URM) bearing wall buildings. Chapter A1 outlines two separate procedures that can be followed to analyze existing URM bearing wall buildings. The first is known as the "general procedure" and is meant specifically for URM buildings that have rigid diaphragms. In truth, most existing URM buildings have flexible wood diaphragms and could therefore not be analyzed in accordance with the general procedure. The second method is termed the "special procedure." The special procedure is based upon research involving URM buildings that were upgraded as part of the Los Angeles hazard reduction program.

In addition to following one of the procedures discussed above, Chapter A1 lists specific requirements for materials, testing, mortar pointing, and submittal requirements.

10.1.2 Chapter A2: Reinforced concrete and masonry buildings. The worst building type for a seismic event that comes to the mind of most engineers is likely a URM building. While these are not ideal buildings in seismic country, neither are older concrete and masonry buildings that include mild reinforcement. In recent earthquakes throughout the United States, reinforced concrete and masonry buildings that have flexible diaphragms have been prone to significant damage.

The seismic deficiencies for these building types first became apparent when several large tilt-up buildings collapsed during the 1971 San Fernando, California, earthquake. Similar damage was evident in subsequent seismic events. The purpose of Chapter A2 is to improve four deficiencies that were identified during past earthquakes. Those four deficiencies are briefly described below:

- Inadequate roof-to-wall and floor-to-wall connections
- Inadequate girder-to-pilaster connections
- Lack of continuous ties across the diaphragm
- Inadequate lateral load distribution at re-entrant corners

10.1.3 Chapter A3: Wood-frame buildings. The four most common weaknesses in wood-frame buildings throughout the United States include the following items:

- Inadequate foundation sill anchorage
- Inadequate bracing of cripple walls
- Discontinuous or inadequate foundations below exterior walls
- Weak or deteriorated masonry or stone foundations

Chapter A3 addresses how a building owner can retrofit their facility to address the sill anchorage and cripple wall bracing deficiencies but does not address inadequate foundations (the last two items noted above). The 2015 IEBC Commentary notes that after the 1989 Loma Prieta earthquake, it was estimated that repairing each of the items listed above would cost on the order of $25,000 to $30,000. That was in 1990 dollars but using the consumer

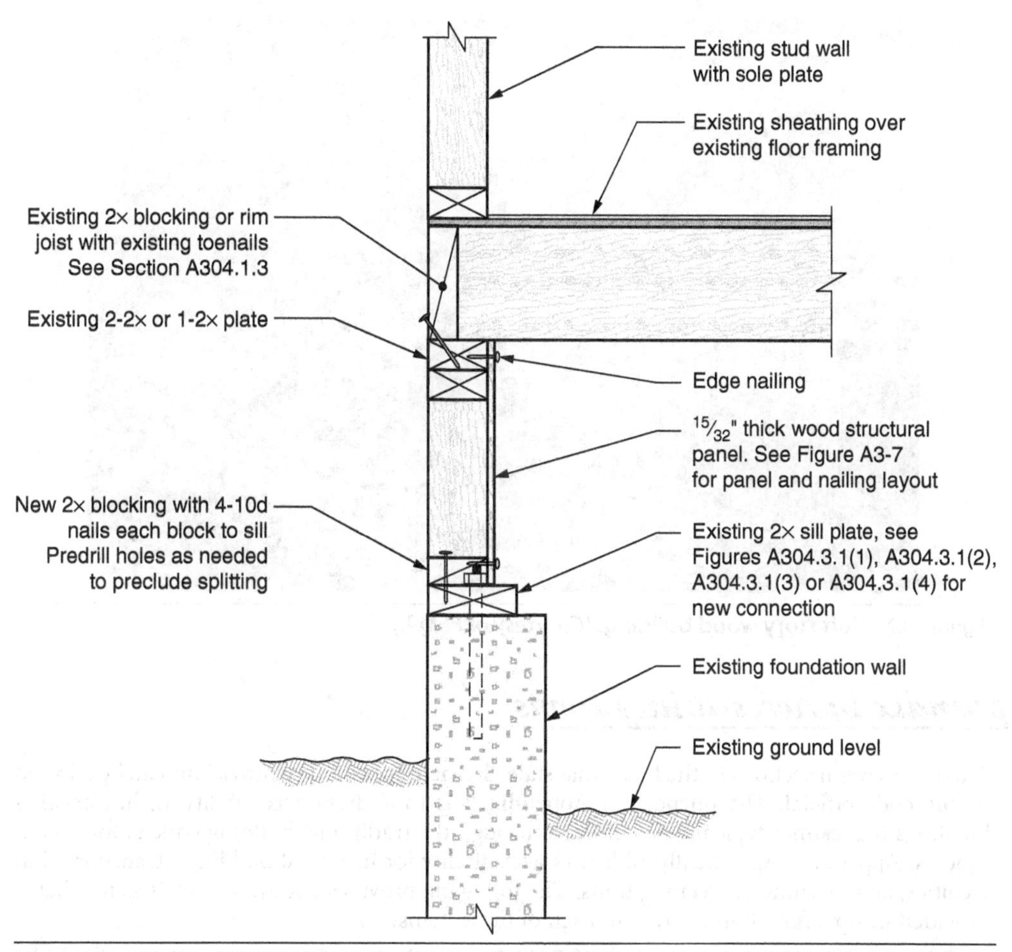

Figure 10.1 Figure A304.41(2) of the IEBC—cripple wall strengthening. (*Excerpted from the 2018 International Existing Building Code; copyright 2017. Washington, D.C.: International Code Council.*)

price index (CPI) inflation calculator provides a cost of $48,000 to $58,000 in 2018 dollars. Figure 10.1 shows how bracing can be provided to an existing cripple wall.

10.1.4 Chapter A4: Open-front wood-frame buildings. Chapter A4 of the IEBC applies to existing wood-framed residential buildings that have soft, weak, or open front walls. These types of buildings are quite common in older areas of California. An example is shown in Fig. 10.2, which is a famous image of a damaged building in San Francisco during the 1989 Loma Prieta earthquake. A soft story building is one in which the first story is substantially weaker and more flexible than the stories above. In Fig. 10.2, large garage openings were provided at the first floor, which made that floor significantly weaker than the floors above. Many jurisdictions in California now have mandatory upgrade requirements for soft story wood buildings. While Chapter A4 provides a method of analyzing and retrofitting these types of buildings, there are several other resources available to building owners and designers that can be used to perform voluntary upgrades to such buildings. In addition, a performance-based analysis approach has been developed by the National Science Foundation and is available at www.nees.org.

Figure 10.2 Soft story wood building. (*Courtesy of FEMA.*)

10.2 *Appendix B: Accessibility Items*

There are several sections in the IEBC that state alternate accessibility provisions can be allowed by the code official. The purpose of Appendix B is to address accessibility within existing buildings that cannot typically be enforced through the traditional building code enforcement process. Appendix B specifically addresses accessibility for historical buildings, transportation facilities, and dwelling or sleeping units. The following provides additional detail as to what is provided in Appendix B in relation to each of these items.

10.2.1 Historical buildings. Section B101.3 states that accessibility compliance should be provided throughout unless the state historic preservation officer determines that compliance would damage the historic significance of the facility. If that is the case, alternative requirements can be provided.

10.2.2 Transportation facilities. Section B102 of the IEBC outlines accessibility requirements for existing transportation facilities. The specific items addressed in this section pertain to existing key stations, accessible routes, and platform and vehicle floor coordination areas. A key station, as outlined in 49 CFR Part 37, includes passenger boarding areas as shown in Fig. 10.3.

10.2.3 Dwelling units and sleeping units. This section simply states that any altered or added dwelling or sleeping units are required to meet the accessibility provisions until the minimum number of accessible units complies with the *International Building Code*® (IBC®).

10.3 *Appendix C: Wind Retrofit*

Appendix C of the IEBC provides retrofit procedures for buildings in high-wind areas. This appendix is broken into two separate chapters. Chapter C1 addresses the proper retrofit of gable ends, while Chapter C2 addresses how roof decks should be fastened in high-wind regions.

Figure 10.3 Washington, D.C., Metrorail station.

It is important to note that there are not any triggers within the IEBC to perform these tasks. Following these procedures is strictly voluntary, but will better protect the public and reduce damage due to large wind events if followed.

10.3.1 Gable end retrofit. Chapter C1 addresses how gable ends can be properly braced such that the structure can better resist large wind loads. A gable end is the portion of a wall that encloses the end of a pitched roof, shown hatched in Fig. 10.4. Not only does the gable end receive a large portion of the wind force, it serves as the main component to transfer the roof loads down to the bearing walls below. They are highly susceptible to damage during large wind events and therefore require adequate bracing.

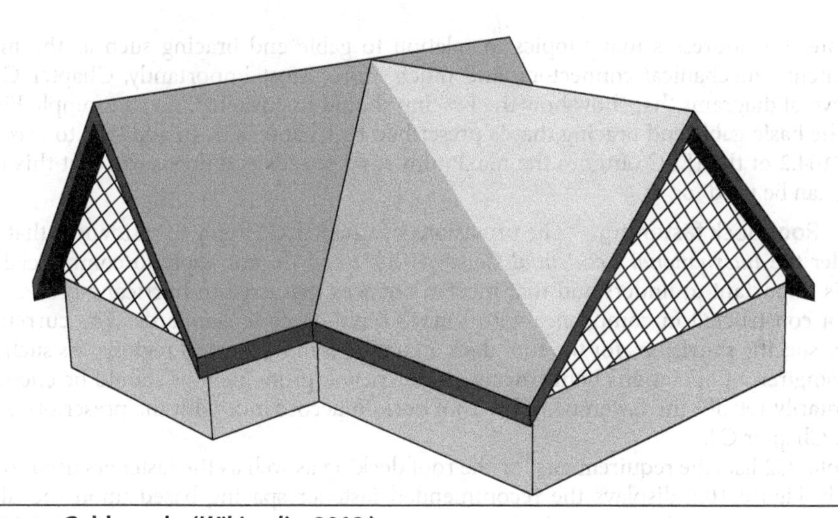

Figure 10.4 Gable ends. (*Wikipedia, 2012.*)

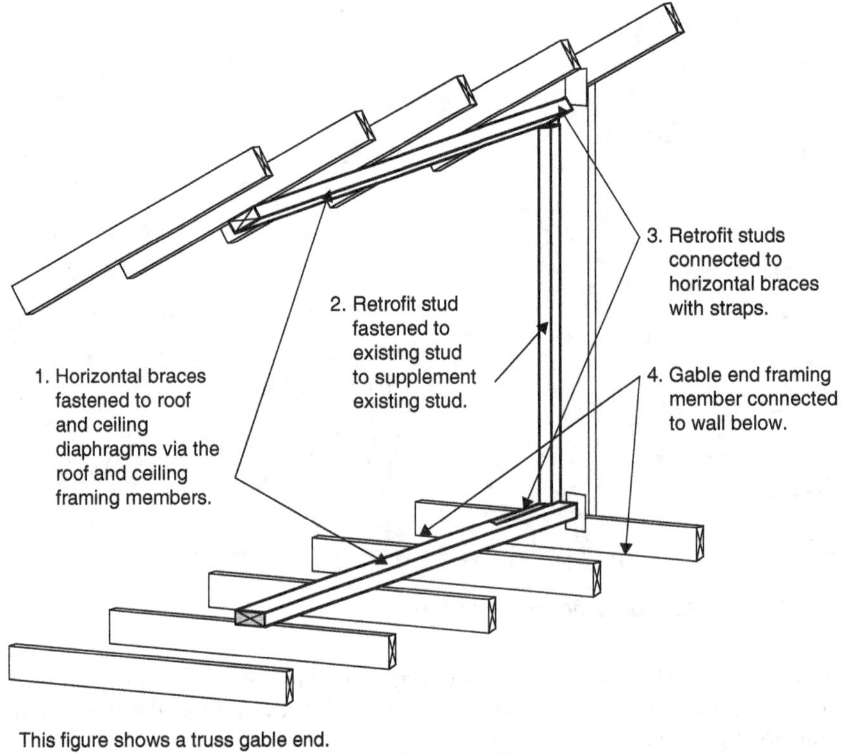
3. Retrofit studs connected to horizontal braces with straps.
2. Retrofit stud fastened to existing stud to supplement existing stud.
1. Horizontal braces fastened to roof and ceiling diaphragms via the roof and ceiling framing members.
4. Gable end framing member connected to wall below.
This figure shows a truss gable end.
The methodology for a conventionally framed gable end is similar.
The numbers indicate a typical sequence of installation.
In order to show straps compression blocks are not shown.

Figure 10.5 Figure C104.1 of the IEBC—gable end bracing. ***(Excerpted from the 2018 International Existing Building Code; copyright 2017. Washington, D.C.: International Code Council.)***

Chapter C1 addresses many topics in relation to gable end bracing such as the material requirements, mechanical connectors, and much more. Most importantly, Chapter C1 provides several diagrams that show how the bracing should be installed. As an example Fig. 10.5 shows the basic gable end bracing that is prescribed by Chapter C1. In addition to this figure, Table C104.2 of the IEBC outlines the maximum wind speeds and exposures that this type of bracing can be used for.

10.3.2 Roof deck fastening. The provisions of Chapter C2 apply to residences that would fall under the *International Residential Code*® (IRC®) and do not apply to commercial facilities. It is important to understand that most residences that exist in hurricane-prone regions were not constructed in compliance with today's building code standards. The current code provides specific provisions for the roof deck attachment in high-wind regions. As such, when undergoing reroof operations home owners in hurricane-prone regions should be encouraged to voluntarily retrofit the fastening of the roof deck in accordance with the prescriptive provisions of Chapter C2.

Chapter C2 lists the requirements for the roof decking as well as the fasteners used to attach the deck. Figure 10.6 displays the recommended fastener spacing based upon the ultimate design wind speed for the project site.

SUPPLEMENT FASTENERS AT PANEL EDGES AND INTERMEDIATE FRAMING				
EXISTING FASTENERS	EXISTING FASTENER SPACING (EDGE OR INTERMEDIATE SUPPORTS	MAXIMUM SUPPLEMENTAL FASTENER SPACING FOR 130 MPH < V_{ult} ≤ 140 MPH	MAXIMUM SUPPLEMENTAL FASTENER SPACING FOR INTERIOR ZONE[c] LOCATIONS FOR MPH V_{ult} > 140 MPH AND EDGE ZONES NOT COVERED BY THE COLUMN TO THE RIGHT	EDGE ZONE[d] FOR V_{ult} > 160 MPH AND EXPOSURE C, OR V_{ult} > 180 MPH AND EXPOSURE B
Staples or 6d	Any	6" o.c.[b]	6" o.c.[b]	4" o.c.[b] at panel edges and 4" o.c.[b] at intermediate supports
8d clipped head or round head smooth shank	6" o.c. or less	None necessary	None necessary along edges of panels but 6" o.c.[b] at intermediate supports of panel	4" o.c.[a] at panel edges and 4" o.c.[a] at intermediate supports
8d clipped head or round head ring shank	6"o.c. or less	None necessary	None necessary	4" o.c.[a] at panel edges and 4" o.c.[a] at intermediate supports
8d clipped head or round head smooth shank	Greater than 6" o.c.	6" o.c.[a]	6" o.c.[a] along panel edges and 6" o.c.[b] at intermediate supports of panel	4" o.c.[a] at panel edges and 4" o.c.[a] at intermediate supports
8d clipped head or round head ring shank	Greater than 6" o.c.	6" o.c.[a]	6" o.c.[a]	4" o.c.[a] at panel edges and 4" o.c.[a] at intermediate supports

For SI: 1 inch = 25.4 mm; 1 foot = 304.8 mm; 1 mile per hour = 0.447 m/s.
a. Maximum spacing determined based on existing fasteners and supplemental fasteners.
b. Maximum spacing determined based on supplemental fasteners only.
c. Interior zone = sheathing that is not located within 4 feet of the perimeter edge of the roof or within 4 feet of each side of a ridge.
d. Edge zone = sheathing that is located within 4 feet of the perimeter edge of the roof and within 4 feet of each side of a ridge.

Figure 10.6 Table C202.1.2 of the IEBC. (***Excerpted from the 2018 International Existing Building Code; copyright 2017. Washington, D.C.: International Code Council.***)

10.4 *Resource A: Archaic Materials*

The fire-resistance ratings of building materials is typically provided by a listed assembly (e.g., UL LLC, Gypsum Association, etc.), by means of the prescriptive requirements in Section 721 of the IBC, or by calculating the rating in accordance with Section 722 of the IBC. While that is the case for new construction, many existing buildings utilize older materials that have not been considered in any listings, nor are the calculated values included under Section 722 of the IBC. For that reason, Resource A has been added to the IEBC to assist the design professional in evaluating the fire-resistance rating of archaic building materials that could be found in existing buildings.

Resource A walks the design professional through how to perform a preliminary evaluation of the building, analyze the effects of penetrations, perform a final evaluation, and then develop a design solution. The resource allows the user to follow an experimental or theoretical approach and provides examples to assist in the process.

REFERENCES

American Concrete Institute (ACI). 2014. "Building Code Requirements for Structural Concrete." ACI 318-14. American Concrete Institute.

American Society of Civil Engineers (ASCE). 2013. "Seismic Evaluation and Retrofit of Existing Buildings." ASCE 41-13. American Society of Civil Engineers.

American Society of Civil Engineers (ASCE). 2016. "Minimum Design Loads and Associated Criteria for Buildings and Other Structures." ASCE 7-16. American Society of Civil Engineers.

Anchor: Modular Buildings. n.d. "Modular Classrooms." Accessed April 1, 2019, https://www.anchormodular.com/modular_classrooms/

ANSI. 2015. "Standards Aid Accessibility on 25th Anniversary of the Americans with Disabilities Act." ANSI website, published August 5, 2015. Accessed October 7, 2018, https://www.ansi.org/news_publications/news_story?menuid=7&articleid=d61afcf2-8560-45c0-998c-7a8d7985fb85

Augenstein, Neal. 2016. "Water Damage Led to Partial Collapse of Fairfax County Condo Building." *Washington's Top News (WTOP)*, October 3, 2016. Accessed September 20, 2018, https://wtop.com/fairfax-county/2016/10/columns-not-ok-in-evacuated-fairfax-co-apartment-building/slide/1/

Baldassarra, Carl F. 2014. "Safety Feature/Safety Hazard." National Fire Protection Agency (NFPA), *NFPA Journal*, September–October 2014. Accessed October 27, 2018, https://www.nfpa.org/News-and-Research/Publications/NFPA-Journal/2014/September-October-2014/Features/Fire-Escapes

Baldwin, Cindy. 2016. "Hazardous Building Materials 101." *Restoration and Remediation Magazine*, August 8, 2016. Accessed September 20, 2018, https://www.randrmagonline.com/articles/87019-hazardous-building-materials-101

Basham, Kim. 2013. ForConstructionPros.com. "Avoid the False Alarm for Low-Strength Concrete." Accessed April 1, 2019, https://www.forconstructionpros.com/concrete/equipment-products/technology-services/article/10846913/avoid-the-false-alarm-for-lowstrength-concrete

BBC, Radio Devon. 2010. "In Pictures: Dartmouth after the Fire." Accessed April 1, 2019, http://kochhars.com/local/devon/hi/people_and_places/newsid_8806000/8806229.stm

Big D Construction. n.d. Accessed April 1, 2019, http://big-d.com/projects/slcc-center-for-new-media-2/

Bonowitz, David. 2017. "What I the Performance Method Trying to Do?" *Structure Magazine*, September 2017, pp. 43–44.

British History Online. n.d. "Charles II, 1666: An Act for Rebuilding the City of London." In *Statutes of the Realm: Volume 5*, 1628-80, ed. John Raithby (s.l: Great Britain Record Commission, 1819), 603–612. Accessed August 18, 2018, http://www.british-history.ac.uk/statutes-realm/vol5/pp603-612

Building Code Development Committee (BCDC). 2017. "Applying Building Codes to Tiny Homes." National Fire Protection Agency (NFPA), BCDC White Paper, May 2017.

Buildinghow, n.d. "Foundation Consisting of Flexible Spread Footings and Connecting Beams." Accessed April 1, 2019, https://www.buildinghow.com/en-us/Products/Books/Volume-A/The-structural-frame/Structural-frame-elements/Foundation

Chicago Architecture Center. n.d. "The Great Chicago Fire of 1871." Accessed August 18, 2018, http://www.architecture.org/learn/resources/architecture-dictionary/entry/the-great-chicago-fire-of-1871/

City of Albany. n.d. "Construction Requirements in the Special Flood Hazard Area (A and AE Zones)." City of Albany, GA, Planning and Development Services. Accessed September 16, 2018, http://www.albanyga.gov/home/showdocument?id=394

City of Lincoln. n.d. "Frequently Asked Questions." City of Lincoln, NE, Building and Safety Department. Accessed September 15, 2018, https://lincoln.ne.gov/city/build/faq.htm

City of Los Angeles. 2014. "Relocation of Buildings." City of Los Angeles, CA, Department of Building and Safety, Information Bulletin, Revised May 24, 2016. Accessed September 17, 2018, http://www.ladbs.org/docs/default-source/publications/information-bulletins/building-code/relocation-of-building-ib-p-bc2014-099.pdf?sfvrsn=10

City of Mercer Island, WA. January, 2013. "Submittal Checklist for Commercial Projects." *WAMU*. Accessed August 25, 2018, http://www.mercergov.org/files/SubmittalCOMM-fillin.pdf

Clark County, Nevada. 2017. "Structural Observation Report." Department of Building and Fire Prevention. Accessed September 1, 2018, http://www.clarkcountynv.gov/building/Forms/Form802.pdf#search=structural%20observation

Colorado Chapter (CCICC). 2016. "Single Family Residential One Story Detached Garage." Colorado Chapter of the International Code Council, December 29, 2016. Accessed September 14, 2018, https://s3-us-west-2.amazonaws.com/coloradochaptericc.org/uploads/2017/07/2015-Garage-fillable-CCICC.pdf

Cote, Arthur E., and Casey C. Grant. 2008. "Codes and Standards for the Built Environment." In *20th Edition of the Fire Protection Handbook*, 1-52 thru 1-63. National Fire Protection Agency.

County of Santa Cruz. 1983. "Relocation of Habitable Structures." County of Santa Clara, CA, Ordinance 12.06.050, August 23, 1983. Accessed September 17, 2018, https://www.codepublishing.com/CA/SantaCruzCounty/html/SantaCruzCounty12/SantaCruzCounty1206.html

Covering, The Corridor. 2016. "Monday Notes: River Tower Updates, Volunteer Recognized, Indigenous People's Day." Accessed April 1, 2019, https://coveringthecorridor.com/2016/11/monday-notes-river-towers-update-volunteer-recognized-indigenous-peoples-day/

Culture Victoria. n.d. "Re-mortaring with Lime Mortar." Accessed April 1, 2019, https://cv.vic.gov.au/stories/built-environment/restoring-st-nicholas/re-mortaring-with-lime-mortar/

Currey, Melody A., and Joseph V. Cassidy. 2016. "2016 Connecticut State Building Code." State of Connecticut, Division of Construction Services, Office of the State Building Inspector. Accessed September 2, 2018, https://portal.ct.gov/-/media/DAS/Office-of-State-Building-Inspector/2016_connecticut_state_building_code.pdf?la=en

Dal Pino, John. 2016. "Seismic Retrofits Using the IEBC." *Structure Magazine*, August 2016, pp. 54–55.

Doughton, Sandi, and Beena Raghavendran. 2015. "Seattle's Old Brick Buildings Could See Huge Damage in Big Quake." *The Seattle Times*, August 10, 2015. Accessed September 1, 2018, https://www.seattletimes.com/business/real-estate/citys-old-brick-buildings-could-see-huge-damage-in-big-quake/

Drucker, Ali. 2015. "17 Tiny Dream Homes Under 200 Square Feet." Trillist. Accessed April 1, 2019, https://www.thrillist.com/home/best-tiny-houses-coolest-tiny-homes-on-wheels-micro-house-plans

DWM. 2012. "Code Update." *Door & Window Manufacturer Magazine*, 13 (5): 42–43, June 2012. Accessed October 27, 2018, https://www.dwmmag.com/digital/2012/Jun2012.pdf

Elcosh. 2014. "Job-built Extinguisher Holder", Electronic Library of Construction Occupational Safety & Health (elcosh). Accessed April 1, 2019, http://www.elcosh.org/image/3861/i002639/Job-built%2Bextinguisher%2Bholder.html

Enervex, Inc. 2011. *Hearth Venting Manual—Fourth Edition*. ENERVEX, Inc. Accessed September 29, 2018, http://www.enervexfansource.com/media/HearthVentingManualEFSv.pdf

Eschenasy, Dan, Gus Sirakis, and Bharat Gami. 2016. "Protection of Existing Buildings During Construction." *2016 Build Safe Live Safe Conference*. Accessed September 1, 2018, https://www1.nyc.gov/assets/buildings/pdf/protection_of_existing_buildings.pdf

Federal Emergency Management Agency (FEMA 480). February 2005. "Floodplain Management Requirements—A Study Guide and Desk Reference for Local Officials." National Flood Insurance Program (NFIP), FEMA 480, February 2005.

Federal Emergency Management Agency (FEMA 547). October 2006. "Techniques for the Seismic Rehabilitation of Existing Buildings." FEMA 547, October 2006.

Federal Emergency Management Agency (FEMA 445). 2006. "Next-Generation Performance-Based Seismic Design Guidelines, Program Plan for New and Existing Buildings." FEMA 445, August 2012.

Federal Emergency Management Agency (FEMA P-762). February 2009. "Local Officials Guide for Coastal Construction." Risk Management Series, FEMA P-762, February 2009.

Federal Emergency Management Agency (FEMA P-758). May 2010. "Substantial Improvement/Substantial Damage Desk Reference." FEMA P-758, May 2010.

Federal Emergency Management Agency (FEMA P-424). December 2010. "Design Guide for Improving School Safety in Earthquakes, Floods, and High Winds." Risk Management Series, FEMA P-424, December 2010.

Federal Emergency Management Agency (FEMA P-499). December 2010. "Home Builder's Guide to Coastal Construction." Risk Management Series, FEMA P-499, December 2010.

Federal Emergency Management Agency (FEMA P-804). December 2010. "Wind Retrofit Guide for Residential Buildings." FEMA P-804, December 2010.

Federal Emergency Management Agency (FEMA P-58-2). September 2012. "Seismic Performance Assessment of Buildings, Volume 2—Implementation Guide." FEMA P-58-2, September 2012.

Federal Emergency Management Agency (FEMA E-74). December 2012. "Reducing the Risk of Nonstructural Earthquake Damage—A Practical Guide." FEMA E-74, December 2012.

Federal Emergency Management Agency (FEMA P-312). June 2014. "Homeowner's Guide to Retrofitting 3rd Edition." FEMA P-312, June 2014.

Federal Emergency Management Agency (FEMA). April 2017. "Understanding Substantial Damage in the International Building Code, International Existing Building Code, or International Residential Code." FEMA Job Aid, April 2017.

Federal Emergency Management Agency (FEMA). April 2017. "Understanding Substantial Structural Damage in the International Existing Building Code." FEMA, April 2017. Accessed August 25, 2018, https://www.fema.gov/media-library-data/1500551566482-21556ab3812be7da04e28f3523595cae/PA_Job-Aid-Understanding_SD-rev.pdf

Federal Emergency Management Agency (FEMA). December 2017. "Flood Zones—Definition/Description." FEMA, December 2017. Updated 12/13/2017. Accessed September 3, 2018, https://www.fema.gov/flood-zones

FHWA. 2005. "NHI Course No. 132078—Micropile Design and Construction Reference Manual." FHWA NHI-05-039 (1-4), December 2005.

Firestop, Southwest. n.d. "Firestopping." Accessed April 1, 2019, http://www.firestopsouthwest.com/firestopping

Fixsproject.com. n.d. "Lateral Loads Seismic Lateral Earth Pressure on Basement and Retaining Walls." Accessed April 1, 2019, http://www.fixsproject.com/lateral-loads/lateral-loads-seismic-lateral-earth-pressure-on-basement-and-retaining-walls/

FloodMaps.com. n.d. "Zone Classifications." Accessed September 3, 2018, http://www.floodmaps.com/zones.htm

Foliente, Greg C. 2000. "Developments in Performance-Based Building Codes and Standards." *Forest Products Journal,* July 2000, Volume 50, Pages 12–21.

ForConstructionPros.com. 2015. "Building Deep Foundations." Accessed April 1, 2019, https://www.forconstructionpros.com/concrete/equipment-products/article/12079396/building-deep-foundations

Formisano, Bob. 2018. "A Visual Guide to Fuses and How They Work." Updated 6/9/2018. Accessed September 29, 2018, https://www.thespruce.com/home-fuse-box-how-they-work-1824667

Gontram Architecture. 2015. "Accessibility: The 20% Rule Explained." Gontram Architecture, posted on June 10, 2015. Accessed October 20, 2018, https://gontramarchitecture.com/accessibility-the-20-rule-explained/

Heady, Eugene J. n.d. "Construction Law—The History Is Ancient." Accessed August 18, 2018, https://www.lexology.com/library/detail.aspx?g=5181e80b-f307-42e6-a357-c2d081b678ff

Hubbell Incorporated. n.d. "Tamper-Resistant and Weather-resistant Receptacles—2008 NEC Code Requirements—What You Need to Know." Accessed September 28, 2018, http://ecatalog.hubbell-wiring.com/press/pdfs/H5272.pdf

Inhabitat. 2016. "Rammed Earth and Weathered Steel Tie Arizona's Maricopa Campus to Native American and Agricultural Roots." Accessed April 1, 2019, https://inhabitat.com/rammed-earth-and-weathered-steel-tie-arizonas-maricopa-campus-to-native-american-and-agricultural-roots/central-arizona-college-by-smithgroupjjr-17/

International Code Council. (n.d.-a). "International Codes—Adoption by State (May 2018)." Accessed August 19, 2018, https://www.iccsafe.org/wp-content/uploads/Code_Adoption_Maps.pdf

International Code Council. (n.d.-b). "International Existing Building Code—Adoption Map." Accessed August 19, 2018, https://www.iccsafe.org/wp-content/uploads/Code_Adoption_Maps.pdf

Keller, Michael, Mira Rojanasakul, David Ingold, Christopher Flavelle, and Brittany Harris. 2017. "Outdated and Unreliable: FEMA's Fault Flood Maps Put Homeowners at Risk." Bloomberg, October 6, 2017. Accessed September 3, 2018, https://www.bloomberg.com/graphics/2017-fema-faulty-flood-maps/

Kilsheimer, Allyn E. 2017. "River Towers Condominiums (Buildings 6621, 6631, and 6641)—In-Progress Forensic Investigation Process—Condition of Exposed and Visible Structural Frame Members Post-October 2016 Collapse." KCE Structural Engineers, P.C. March 13, 2017. Accessed September 20, 2018, https://static1.squarespace.com/static/53626999e4b0ec04b0ff5766/t/58c9a15d2994ca9063514543/1489609053753/2016-18+-+River+Towers+In-Progress+Investigation+Report+FINAL.pdf

Laslo, Matt. May 20, 2011. "Bus Crashes into KFC in Silver Spring." *WAMU*. Accessed August 25, 2018, https://wamu.org/story/11/05/20/bus_crashes_into_kfc_in_silver_spring/

Mackey, Brian. 2016. "Legislators Weigh Pension Buyouts." NPR Illinois. Accessed April 1, 2019, HYPERLINK "http://www.nprillinois.org/post/legislators-weigh-pension-buyouts" http://www.nprillinois.org/post/legislators-weigh-pension-buyouts#stream/0

Maison, Bruce, Russell Berkowitz, Heidi Faison, Justin M. Spivery, and Mohamed Talaat. 2010. "Perspectives on ASCE 41 for Seismic Rehabilitation of Buildings." *Structure Magazine*, October 2010, p 26.

Martin, Zeno, Brian Tognetti, and Howard Hill. 2017. "Current Code and Repair of Damaged Buildings." *Structure Magazine*, February 2017, pp. 22–24.

McCormick, David L. 2003. "Seismic Retrofit—Strengthening Tilt-up Structures." *Structure Magazine*, August/September 2003, pp. 1–4.

Orenstein, Natalie. 2017. "Water-Absorbent Material Caused Fatal Balcony Collapse, State Says." Berkelyside.com, June 2, 2017. Accessed September 1, 2018, https://www.berkeleyside.com/2017/06/02/water-absorbent-material-caused-fatal-balcony-collapse-state-says

Pacific Earthquake Engineering Center (PEER). 2017. "Guidelines for Performance-Based Seismic Design of Tall Buildings." PEER, Tall Buildings Initiative. Version 2.03, May 2017.

Rhoads, Marcela Abadi. 2016. "ANSI vs. ADA." Abadi Accessibility, posted on January 4, 2016. Accessed October 6, 2018, http://www.abadiaccess.com/uncategorized/ansi-vs-ada/

Roberts, Nicole. 2018. "City Officials Warn against Dangerous Buildings." *News Tribune*, March 4, 2018. Accessed September 2, 2018, http://www.newstribune.com/news/news/story/2018/mar/04/city-officials-warn-against-dangerous-buildings/715983/

Rodgers, Emory, Sara Yerkes, John Terry, Steve Orlowski, and Paul Karrer. 2017. "The Role of Existing Building Codes in Safely, Cost-Effectively Transforming the Nation's Building Stock." A White Paper by the National Institute of Building Sciences, National Council of Governments on Building Codes and Standards (NCGBCS).

Rojahn, Christopher. 2014. "Technical Focus: The ATC-58 Project." *Civil + Structural Engineer Magazine*, February 19, 2014. Accessed September 23, 2018, https://csengineermag.com/article/technical-focus-the-atc-58-project/

Schons, Mary. 2011. "The Chicago Fire of 1871 and the 'Great Rebuilding.'" Accessed August 18, 2018, https://www.nationalgeographic.org/news/chicago-fire-1871-and-great-rebuilding/

Scott, Donald R. 2018. "ASCE 7-17 Wind Load Provisions." *Structure Magazine*, July 2018, pp. 12–14.

SEAU.org. 2018. "Re-Roofing Questionnaire." Structural Engineer's Association of Utah (SEAU), issued July 10, 2018. Accessed October 23, 2018: https://www.seau.org/library/CB82C28F-2AF1-0A7B-2065-73715B034A34

Smith, Nathaniel, and Milan Vatovec. 2015. "Mitigating and Remediating Damage to Properties Adjacent to Construction in Congested Urban Environments." ICRI Spring 2015 Convention. Accessed September 1, 2018, https://c.ymcdn.com/sites/icri.site-ym.com/resource/resmgr/Events/07_Smith.pdf

Spokane County. 2011. "Relocation Permits—Residential." Spokane County, WA, Department of Building and Planning. Accessed September 17, 2018, https://www.spokanecounty.org/DocumentCenter/View/635/BP-05-Relocation-Permits-PDF

State of Michigan. 1972. "Construction Board of Appeals: Creation, Appointment, Qualifications, and Terms of Members." Section 125.1514 of Stille-Derossett-Hale Single State Construction Code Act. Accessed September 2, 2018, http://www.legislature.mi.gov/(S(kl1elvw1inaw1dv53gcbidcx))/mileg.aspx?page=getObject&objectName=mcl-125-1514

Superior Awning. n.d. "Mobile Home Shade Covers." Accessed April 1, 2019, https://superiorawning.com/mobile-home-awnings/

Systems, Universal Scaffold. n.d. "Sidewalk Protection Scaffold Rentals." Accessed April 1, 2019, https://www.universalscaffoldrentals.com/sidewalk-protection-scaffold-rental/

U.S. Access Board, 2014. "ADA Scoping: Alterations and Additions." United States Access Board Technical Guide, February 2014. Accessed October 20, 2018, https://www.access-board.gov/attachments/article/999/alterations.pdf

USSC. 2016. *The Utah Guide for the Seismic Improvement of Unreinforced Masonry Dwellings.* 2nd ed. Utah Seismic Safety Commission (USSC), 2016.

WBDG Accessible Committee. 2017. "History of Accessible Facility Design." Whole Building Design Guide (WBDG), a Program of the National Institute of Building Sciences, Accessible Committee. Updated October 30, 2017. Accessed October 6, 2018, https://www.wbdg.org/design-objectives/accessible/history-accessible-facility-design

Whatcom County, WA. n.d. "Site Plan Requirements." Whatcom County, Planning and Development Services. Accessed September 14, 2018, https://www.whatcomcounty.us/DocumentCenter/View/2094/Site-Plan-Requirements-PDF?bidId=

Wikimedia. 2008. Accessed April 1, 2019, https://commons.wikimedia.org/wiki/File:Rotermann_Quarter.jpg

Wikimedia. 2009. Accessed April 1, 2019, https://commons.wikimedia.org/wiki/File:Salem_Church_Relocation.JPG

Wikimedia. 2010. Accessed April 1, 2019, https://commons.wikimedia.org/wiki/File:Property_damage_at_Diego%27s_Hair_Salon_-_Blizzard_of_2010.JPG

Wikimedia. 2016a. Accessed April 1, 2019, https://commons.wikimedia.org/wiki/File:Chicago_in_Flames_by_Currier_%26_Ives,_1871.jpg

Wikimedia. 2016b. Accessed April 1, 2019, https://commons.wikimedia.org/wiki/File:Tree_on_house,_Hurricane_Hermine,_Valdosta,_Georgia.jpg

Wikipedia. n.d. "International Building Code." Accessed August 18, 2018, https://en.wikipedia.org/wiki/International_Building_Code

Wikipedia. 2012. Accessed April 1, 2019, HYPERLINK "https://en.wikipedia.org/wiki/Gable" https://en.wikipedia.org/wiki/Gable#/media/File:Gables.jpg

Wikiwand. n.d. "Fuse (Electrical)." Accessed April 1, 2019, http://www.wikiwand.com/en/Fuse_(electrical)

Wisconsin State Legislature. 2018. "Chapter SPS 366—Existing Buildings." Department of Safety and Professional Services (SPS), Commercial Building Code, April 2018. Accessed September 8, 2018, https://docs.legis.wisconsin.gov/code/admin_code/sps/safety_and_buildings_and_environment/361_366/366/I/1401

Wittenberg, Ariel. 2017. "Extreme Weather—The Myth of the 100-year Flood." *E&E News*, published August 20, 2017. Accessed September 3, 2018, https://www.eenews.net/stories/1060059425

www.thegolfclub.info. n.d. "Code Hammurabi Drawing." Accessed April 1, 2019, http://www.thegolfclub.info/related/code-hammurabi-drawing.html

INDEX

F

G

H

I

L

M

O

P

R

S

T

U

V

W

ICC INTERNATIONAL
CODE
COUNCIL®
People Helping People Build a Safer World®